ATLAS OF HUMAN PRENATAL MORPHOGENESIS

ATLAS OF HUMAN PRENATAL MORPHOGENESIS

JAN E. JIRÁSEK, M.D., D. Sc.

Institute of the Care of Mother and Child, Prague
and
Department of Obstetrics and Gynecology, Medical School
University of Minnesota, Minneapolis

with the technical assistance of
B. FALTINOVÁ and L.R. SWENEY

1983/ **MARTINUS NIJHOFF PUBLISHERS**
a member of the KLUWER ACADEMIC PUBLISHERS GROUP
BOSTON / THE HAGUE / DORDRECHT / LANCASTER

Distributors

for the United States and Canada: Kluwer Boston, Inc., 190 Old Derby Street, Hingham, MA 02043, USA
for all other countries: Kluwer Academic Publishers Group, Distribution Center,P.O.Box 322, 3300 AH Dordrecht, The Netherlands

Library of Congress Cataloging in Publication Data   CIP

```
Jirásek, Jan Evangelista.
   Atlas of human prenatal morphogenesis.

   1. Embryology, Human--Atlases.  I. Faltinová, B.
II. Sweney, L. R.  III. Title.  [DNLM: 1. Embryology
--Atlases.  2. Morphogenesis--Atlases.  QS 617 J61a]
QM602.J57  1983      612'.64      82-12573
```

ISBN-13: 978-94-009-6698-7   e-ISBN-13: 978-94-009-6696-3

DOI: 10.1007/978-94-009-6696-3

This book is dedicated
to my friends from the University of Minnesota.

CONTENTS

PREFACE

A little picture is worth a million words.

Chinese proverb

Prenatal human development is an extremely complicated process related to genetics, biochemistry, anatomy, and physiology. There are no developmental changes, either chemical or morphologic, without simultaneous changes in molecular organization. The astonishing buildup of biostructures always precedes their proper function. The development of an embryo is genetically coded and is based on interactions related to the selective switching on and off of genes. Interactions are cell-to-cell mediated, mediated by extracellular fluids, or mediated by special pathways. Every substance involved in developmental interactions, before triggering a metabolic or a morphogenic event, is to be recognized by its target. Complex physical and immunologic recognitions are involved in the process of differentiation. Small pieces of evidence are collected to create a mosaic picture elucidating the development. This picture is fascinating and represents the biggest biological puzzle: the puzzle of development.

There is no doubt that analysis of human prenatal development is a basis for understanding normal and pathologic relationships between structure and function. Today, there are approximately 2000 different inborn congenital anomalies and syndromes. This book attempts to present a complete realistic account of human morphogenesis, the differentiation of structures, using direct photographs of normal specimens obtained from legal medical abortions of unwanted pregnancies. Emphasis has been placed on contemporary techniques: histochemistry and scanning electron microscopy. The text is as simple as possible; meticulous detailed anatomic descriptions have been omitted.

The basic anatomic contributions to human development were made many years ago. Probably still the best classic description of human organogenesis can be found in the *Manual of Human Embryology* (two volumes) edited by Keibel and Mall in 1910 and 1912. The comprehensive anatomic analysis and staging of human embryos was published by Streeter in his impressive monograph *Developmental Horizons in Human Embryos* (Washington, DC: Carnegie Institute of Embryology, 1951), and by O'Rahilly in *Developmental Stages in Human Embryos* (Washington, DC: Carnegie Institute, 1972). Personally, I admire the information on early human differentiation provided by Hertig and Rock in their papers published from 1941 to 1956 (see O'Rahilly's monograph).

Textbooks of human embryology, such as Patten's *Human Embryology* (New York: McGraw-Hill, 3rd edn., 1968), Davies' *Human Developmental Anatomy* (New York: Ronald Press, 1963), Hamilton, Boyd, and Mossman's *Human Embryology* (Baltimore: William and Wilkins, 4th edn., 1972), and Langman's *Medical Embryology*, Baltimore: Williams and Wilkins, 3rd edn., 1975) should be consulted if, in some cases, the information provided in this atlas seems insufficient. The comparison of illustrations and simplified drawings and schemas of such textbooks with the photographs presented in this book is recommended. Morphogenesis is a three-dimensional process and cannot be satisfactorily illustrated by histologic sections. Therefore this atlas is based on microdissections, scanning electron microscopy, and surface histochemical staining. To present the results based on staining reactions, it was necessary to use color photographs. Their reproduction is quite complicated and expensive, but there was no other choice. Techniques of 'surface histochemistry' and scanning electron microscopy not only produce suprising results, but also make developmental anatomy more closely related to life than do classic wax models and faded specimens preserved in embryologic museums in formalin- or alcohol-filled bottles. While embryology appeared to be a dead science, molecular biologists, biochemists, and experimental embryologists made impressive progress. In most cases, however, there is no conclusive evidence that results related to experimental animals are applicable to man. A recent task has been to bridge the interdisciplinary gaps.

Acknowledgments

I would like to express my deep gratitude to B. Faltinová and L.R. Sweney for their outstanding technical assistance. I also would like to thank to my friends, Professor K.A. Premm, chairman and head of the Department of Obstetrics and Gynecology at the Medical School of the University of Minnesota, who invited me to join his department for the 1978–1979 period; and Professor B.L. Shapiro, who gave me the opportunity to use the scanning microscope at the Division of Oral Biology at the Dental School of the University of Minnesota, Minneapolis. Without their help, this book could not have been written. It has been a great pleasure to work with Martinus Nijhoff Publishers and I am grateful to them for the attention they have given to my book.

Jan E. Jirásek
Prague, Czechoslovakia

MATERIAL AND METHODS

Liberalization of abortion law in Czechoslovakia in 1958 enabled the collection of a reasonable number of normal human embryos that could be used for medical studies. Since 1960, I have collected, dissected, and incubated with various substrates, more than 500 specimens. Clinical history was recorded for all embryos. Embryonic age was calculated from the last normal menstrual period of the mother. The conceptional age was calculated by adding 14 days to the first day of the last menstrual period of the mother if the menstrual cycle was of the regular 28 days' duration. In some specimens, the term of conception was known from the anamnesis.

To achieve the results presented in this book, predominantly two methods were used: surface histochemistry and scanning electron microscopy (SEM). Methods of surface histochemistry are based on our observations that dissected unfixed, or properly fixed, tissue blocks can be subjected to histochemical reactions. There are tridimensional territorial differences in the staining intensity, which can be used to identify special structures or to visualize 'developmental areas and fields'. The impressive combination of alkaline phosphatase stain and alcian-blue glucoseaminoglycan stain was performed in the following way: Embryos obtained from medical abortions were immediately fixed in 10% Ca formol at $2° - 4°$ C for 1 h, washed for 1 h in several changes of cold saline, dissected, and incubated in the following solution: 20 mg naphthyl AS-phosphate dissolved in one drop of dimethyl formamid, 10 ml of 0.1 M veronal acetate buffer pH 8.3, and 40 mg fast-red TR salt. The solution was filtered and the incubation lasted for 1–3 h at 37° C. After incubation, specimens were washed with saline and after-stained in the following solution: 10 ml of 40% formaldehyde, 15 ml of 96% alcohol, 5 ml acetic acid, 20 ml distilled water, and 1 ml of 1% alcian blue in 1% acetic acid. The staining required from 1 h to several days (cartilaginous skeleton). After staining, specimens were washed with water and stored in a solution of 10 ml formaldehyde, 10 ml alcohol, 30 ml water, and 2 ml acetic acid.

By this method, alkaline phosphatase activity appeared red, and proteoglycans and glucoseaminoglycans stained blue. The intensity of staining related to the contents of different glucoseaminoglycans in the tissues was controlled by adjusting the pH of the staining solution. I introduced this method of surface histochemical staining to the United States in 1976, so the method was called the 'bicentennial stain' by some of my American friends.

Specimens for scanning electron microscopy were processed in the following way: Immediately after abortion, embryos were fixed in 10% formol with 1% $CaCl_2$ buffered by excess of $CaCO_3$ at $2° - 4°$ C. After fixation for 1–5 h, specimens were washed in saline and stored in 70% or 80% ethanol for several months. The specimens were dissected under the dissection micriscope and rehydrated through 70%, 50%, and 30% ethanol for 1 h in each solution. They were washed in three changes of 0.1 M cacodylate buffer at pH 7.4 and afterfixed in 3% glutaraldehyde–tannic acid (GTA procedure; L.R. Sweney and B.L. Shapiro, 1977, 'Rapid preparation of uncoated biological specimens for scanning electron microscopy', *Stain Technology* 52: 221, 1977). Between each fixation step, the specimens were thoroughly washed in 0.1 M cacodylate buffer at pH 7.4. After fixation, the specimens were dehydrated through graded ethanols, absolute ethanol, and freon. They were critical-point dried, and mounted on aluminum stubs. All specimens were examined and photographed, without further treatment, in an Etec Autoscan scanning electron microscope (Etec Corporation, Hayward, CA, USA) at 20 kV.

There is considerable shrinkage of the embryonic tissues, but the general morphology remains unchanged and acceptable. There is no doubt that scanning electron microscopy of human embryos contributes to our knowledge of developmental morphologic changes in an unprecedented way.

1. EXTERNAL FORM OF THE EMBRYO AND DEVELOPMENTAL STAGES

Prenatal morphogenesis covers a period starting with fertilization and ending with parturition – expulsion of the fetus from the uterus. At the beginning of this period, the unique genetic system of an individual is established by the union of the male and female germ cells (fertilization). Subsequent development of the fertilized oocyte gives rise to the conceptus: embryo, fetus, newborn, and fetal envelopes. The role of genetic factors and their interactions with environmental factors are largely unknown.

The *prenatal period* of human life comprises about 38 weeks from fertilization until delivery (conceptional age), or about 40 weeks from the first day of the last menstrual period of the cycle in which conception took place (gestational age). For anatomic purposes, usually the conceptional age is used; for clinical purposes, usually gestational age is used. The prenatal period is arbitrarily divided into an embryonic period (gestational weeks 1–10), a fetal period (gestational weeks 11–28), and a perinatal period (gestational weeks 29–40).

During the *embryonic period*, blastogenesis (formation of the germ layers) and early organogenesis (formation of the primordia of organs) occur. The term 'embryo' is used for an organism undergoing blastogenesis and early organogenesis. Fusion of the eyelids is regarded as the arbitrary end of the human embryonic period, which covers conceptional weeks 1–8.

The *fetal period* includes conceptional weeks 9–26. During this period, all organs become functional (sometimes in a different way than postnatally). The fetus is characterized by fused eyelids and its total body weight increases from ca. 2 g to 1000 g.

At the beginning of the *perinatal period*, the eyes of the fetus reopen and, if delivered, the fetus is called a newborn – immature at gestational weeks 28–32, premature at gestational weeks 32–36, and full term from gestational week 36 until week 42. If the gestational period exceeds 42 weeks (more than 40 conceptional weeks), the newborn is called postmature.

Stages of human development

Stage 1: the unicellular stage

The one-cell (oocyte) stage (Figs. 1c–3c*): age 0–24 h after ovulation. The oocyte is expelled from the ovary after the first meiotic division has been completed and the first polar body detached. The oocyte is enclosed within the zona pellucida with some attached residual follicular cells. Fertilization takes place in the infundibulum of the uterine tube. Only one sperm penetrates into the oocyte; the sperm head breaks from the tail and its chromatin expands and transforms into the male pronucleus, During the formation of the male pronucleus, the oocyte undergoes the second meiotic division, resulting in the formation of the second polar body and the female pronucleus. Each pronucleus contains a haploid set of chromosomes. The DNA in each pronucleus is duplicated and consequently the homologous paternal chromosomes of the male pronucleus pair in the equatorial level with maternal chromosomes of the female pronucleus. A unique genome of a new individual becomes established. As new DNA is synthesized in the pronuclei, the haploid amount of DNA of each pronucleus increases into a diploid amount and the total amount of DNA within the fertilized oocyte increases into a tetraploid.

* The letter c after figure numbers means that color figures are being cited. All color figures are presented at the end of this book.

4

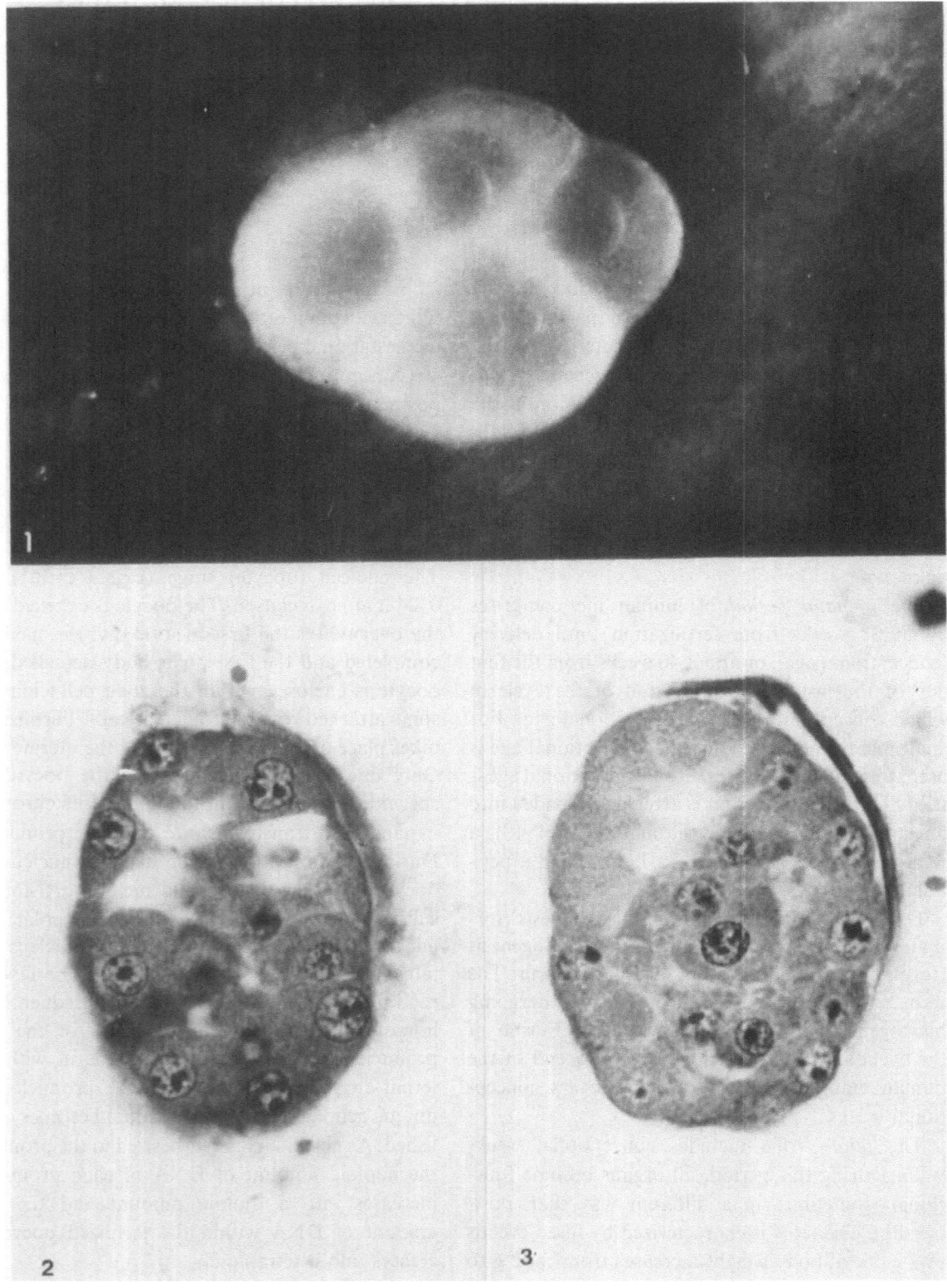

Figures 1–3. (1) Four-cell morula enclosed by zona pellucida, washed out from the tube; ca. 60 h after coitus. Stage 2. **(2)** Early blastocyst. Stage 3. Formation of trophoblast, embryoblast (inner cell mass), and blastocystic cavity. Carnegie Institute, Washington, DC, Department of Embryology, spec. no. 8794; courtesy of Professor O'Rahilly. **(3)** An other section of the same specimen.

Consequently the oocyte divides into two blastomeres by the first mitotic division. Each of the two first blastomeres contains a diploid amount of DNA. The following three *substages* are recognized during the unicellular stage:

1-1 The penetrated oocyte: comprises attachment of the sperm to the oocyte, its penetration into the cytoplasm of the oocyte, and the formation of the male pronucleus.
1-2 The ootid: characterized by the presence of two pronuclei, male and female.
1-3 The zygote: stage of the first mitosis.

Stage 2: the blastomeric stage
The stage of egg segmentation (oocyte) (Fig. 1): age 24–74 h. By sequential mitotic divisions, the fertilized oocyte cleaves into blastomeres, which are enclosed within the zona pellucida. As the number of blastomeres increases, their size decreases. The spherical aggregate of blastomeres, 2–16 in number, enclosed in the zona pellucida is known as the morula, which is transported by the oviduct to the uterotubal junction and enters the uterus ca. 72–74 h after ovulation. *Substaging* of stage 2 is usually related to the number of blastomeres.

Stage 3: the blastodermic stage
The free blastocyst (Figs. 2 and 3): age 74–120 h. The blastocyst is a hollow fluid-filled sphere composed of external and internal blastomeres. As intercellular junctions are formed between outer blastomeres, fluid accumulates underneath the blastomeres, and the morula is converted into the blastocyst. The external blastomeres give rise to a single-layered trophoblast. The internal blastomeres, excentrically apposed to the trophoblast, contribute to the inner cell mass (the embryoblast). The fluid-filled blastocystic cavity is known as the blastocoele. The trophoblast covering the inner cell mass is called polar and that of the 'free' wall of the blastocyst is known as mural. As fluid accumulates within the blastocoele, and as the number of cells increases, the size of the blastocyst increases. The zona pellucida is shed ('hatching of the blastocyst'). Free blastocysts are found within the uterine cavity.

Substages:
3-1 Unhatched blastocyst: blastocyst with the zona pellucida.
3-2 Hatched blastocyst: blastocyst without zona.

Stage 4: the bilaminar embryo stage
The stage of endoderm formation, gastrulation (Figs. 4c and 5c; 4 and 5): age 6–14 days. The inner cell mass of the blastocyst becomes organized into two primary embryonic layers: the ectoderm and the endoderm. Both layers contribute to the bilaminar embryonic disk. Between the ectoderm of the early embryonic disk and the trophoblast, there is a slit-like amniotic cavity. The endoderm is formed by delamination from the ectodermal cells. The ectodermal cells give rise to the amniotic sac and the endodermal cells give rise to the primary yolk sac. The implantation occurs at the early stage of the bilaminar embryo. The blastocyst penetrates into the propria of the endometrium and becomes embedded in its compact superficial layer. Implantation is a trophoblast–endometrium interaction phenomenon beginning with attachment of the trophoblast to the uterine epithelium and ending with the formation of a complete trophoblastic shell. During implantation, the blastocyst collapses, losing fluid, and the single-layered trophoblast of the blastocyst transforms into the trophoblastic shell formed by cyto- and syncytiotrophoblasts.

The blastocyst reexpands after implantation. Irregular spaces within the early trophoblastic shell fuse into a primordium of the intervillous space. Stellate cells appearing underneath the cytotrophoblast covering the inside of the blastocoele constitute the primary mesoderm. Some believe that mesenchymal cells representing the primary mesoderm are detached from the cytotrophoblast, while others consider the primary mesoderm as cells released from the inner cell mass. The superficial cells of the primary mesoderm lining the blastocoele join the endoderm of the primitive embryonic disk and contribute to the extraembryonic portion of the primary yolk sac known as the Heuser's, or exocoelomic, membrane. In this way the primary yolk sac is formed by delaminated endoderm of the embryonic disk and by the Heuser's membrane originating from primary mesoderm.

The trophoblast supported by the mesenchyme (primary mesoderm) is known as the chorion. The chorion grows much faster at this stage than the bivesicular ecto-/endodermal anlage of the embryo.

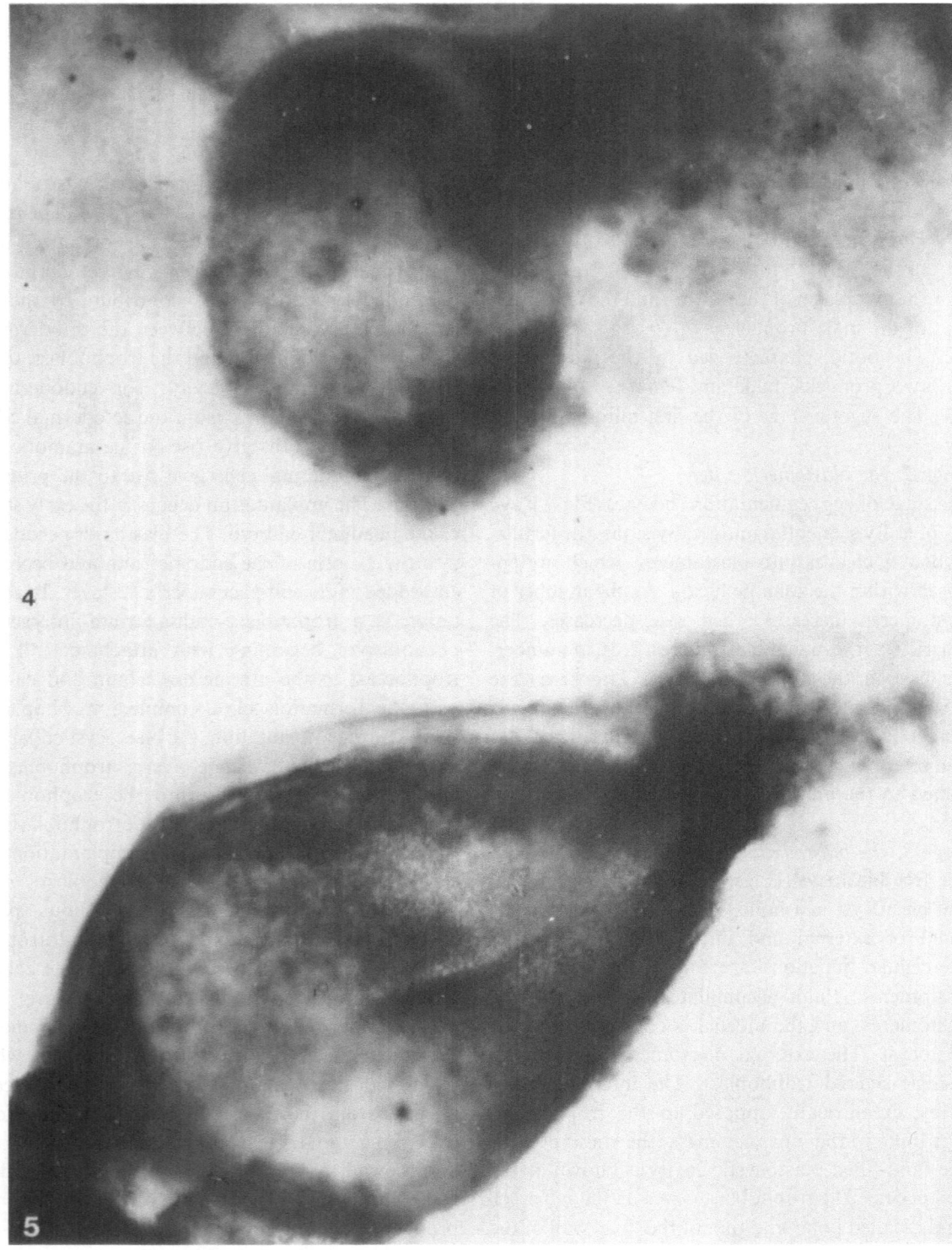

Figures 4 and 5. (4) Unsectioned bilaminar embryo consisting of the amniotic sac and the secondary yolk sac, attached to the chorion by the connecting stalk. Stage 4-3 Lateral view: the embryonic disk is dorsally convex. **(5)** Unsectioned trilaminar embryo. Stage 5-2. The yolk sac was dissected and the endodermal portion of the embryonic disk exposed. The primitive groove is evident.

The mesenchyme penetrates into the cytotrophoblastic columns of the trophoblastic shell, transforming them into the chorionic villi. The size of the blastocystic cavity rapidly increases. Consequently the relatively large primary yolk sac ruptures and closes again, giving rise to a minor secondary yolk sac underneath the embryonic disk and to a residual endodermal cyst adjacent to the chorion.

The two embryonic vesicles – the amniotic sac and the secondary yolk sac – are embedded in a thin sheet of primary mesoderm (the mesenchyme) and are attached to the chorion by a mesenchymal connecting stalk. Insertion of the connecting stalk determines the caudal portion of the embryonic disk.

Substages:

4-1 Bilaminar embryonic disk with incomplete amniotic and yolk sacs.

4-2 Bilaminar embryo with a primary yolk sac.

4-3 Bilaminar embryo with a secondary yolk sac.

The bilaminar embryo consists of a round embryonic disk formed by adjacent round disks of embryonic ectoderm and endoderm interposed between cavities of the amniotic and yolk sacs. The disk is ca. 0.3 mm long and ca. 0.4 mm wide.

Stage 5: the trilaminar presomite stage

The stage of the formation of unsegmented intraembryonic mesoderm and notochord – the stage of the primitive streak and the primitive node (Figs. 6c–12c; 5 and 6): age 15–20 days, embryonic length 0.5–1.5 mm. Trilaminar presomite embryos are characterized by mesoblastic (mesodermal) proliferation from the primitive streak, which is an axial linear thickening of the ectoderm extending from the middle of the embryonic disk to the connecting stalk. The centrally located axial longitudinal linear depression in the primitive streak is known as the primitive groove. The primitive groove brings ectoderm and endoderm into close contact. Cellular condensation with a distinct dimple at the most anterior portion of the primitive streak represents the primitive knot (the Hensen's node). A small area of endoderm anterior to the primitive knot thickens into the prochordal plate. The posterior wall of the yolk sac close to the posterior end of the primitive streak evaginates as a narrow endodermal canal into a connecting stalk

and is known as the allantois. The close apposition of ectoderm and endoderm of the allantois near the opening of allantois into the yolk sac contributes to the bilaminar, ecto-/endodermal, cloacal membrane.

The cells from the primitive streak migrating into the space between ectoderm and endoderm represent the third embryonic layer, the (intraembryonic) mesoblast (mesoderm). The mesoderm contributed to by the primitive streak spreads to all sites, reaches the margin of the embryonic disk, and connects with the sheet of the extraembryonic mesoderm located around the amniotic sac and the yolk sac.

Cells proliferating from the Hensen's node grow anteriorly into the endodermal ceiling of the yolk sac as an axial cellular cord, the chordomesodermal process. As the chordomesodermal process (notochordal canal or head process) elongates, the shallow depression in center of the primitive knot – the primitive pit – deepens and the chordomesodermal process becomes lumenized. The chordomesodermal canal extends from the primitive knot to the prochordal plate. Consequently the ventral wall of the chordomesodermal canal disintegrates, and the canal changes into the notochordal plate, which is incorporated as an axial structure within the ceiling of the yolk sac. A temporary communication between the amniotic cavity and the yolk-sac cavity through the primitive pit is known as the neurenteric canal. The ectoderm of the embryonic disk anterior to the primitive streak and above the notochordal canal contributes to the neural plate (the anlage of the CNS). The direct contact between the notochordal plate and the neural plate determines the formation of the ventral plate of the CNS. Consequently the neural plate changes into the neural groove with elevated neural folds. The formation of the chordomesodermal canal and the chordomesodermal plate is considered as the early notogenesis (neurulation). The embryonic disk at stage 5 is pear shaped.

Substages:

5-1 Trilaminar presomite embryo with a primitive knot and primitive streak.

5-2 trilaminar presomite embryo with a primitive streak and with a distinct notochordal process; neural plate or folds.

8

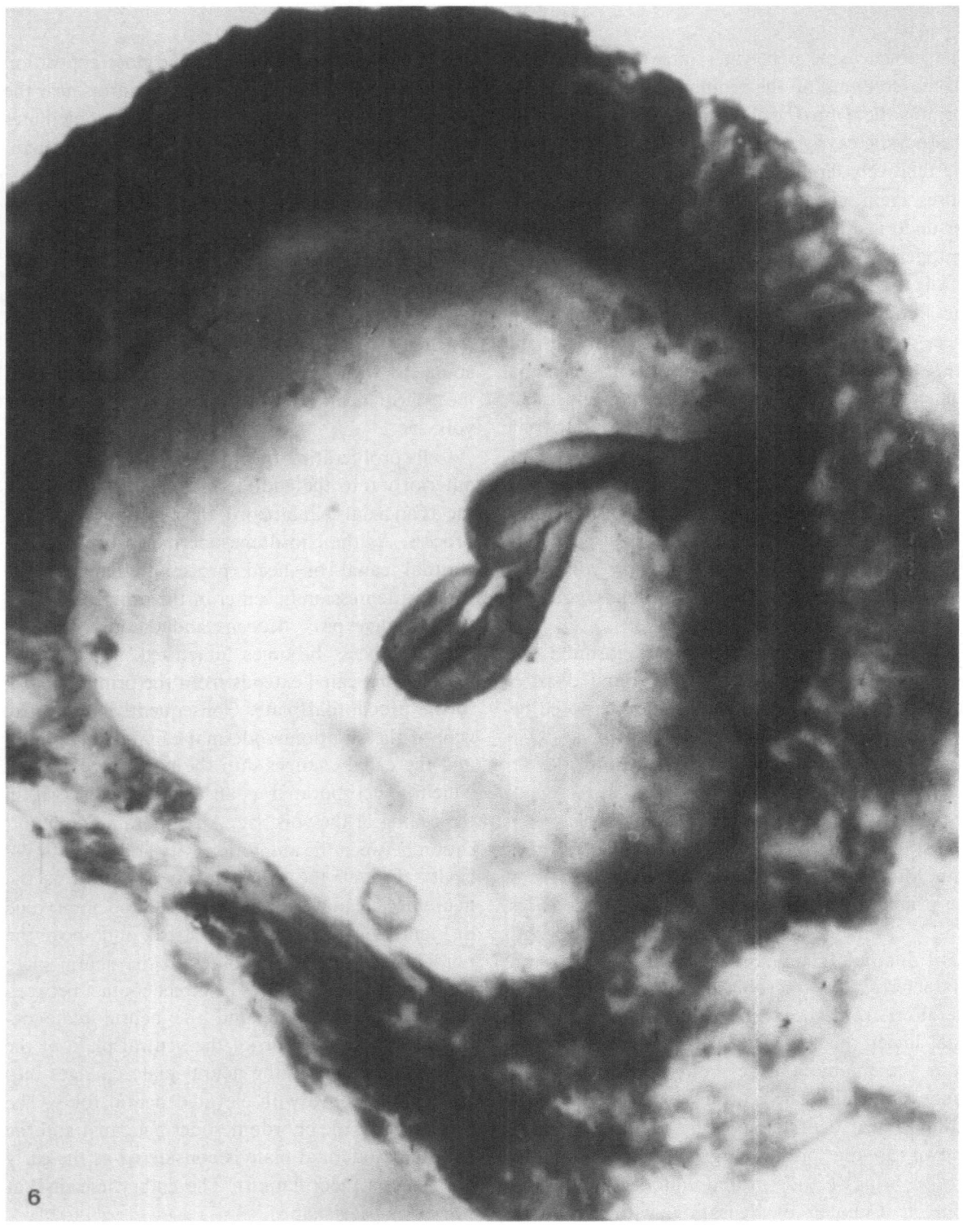

Figure 6. Conceptus. Stage 5-2. The chorionic vesicle is dissected. Lateral view of an unsectioned presomite trilaminar embryo 1.5 mm long, ca. 19–20 days old, with an intact amniotic sac attached to the chorion by the connecting stalk. The yolk sac is partially collapsed and pushed to the left side. An endodermal cyst (bottom) is attached to the chorion. Unfixed specimen.

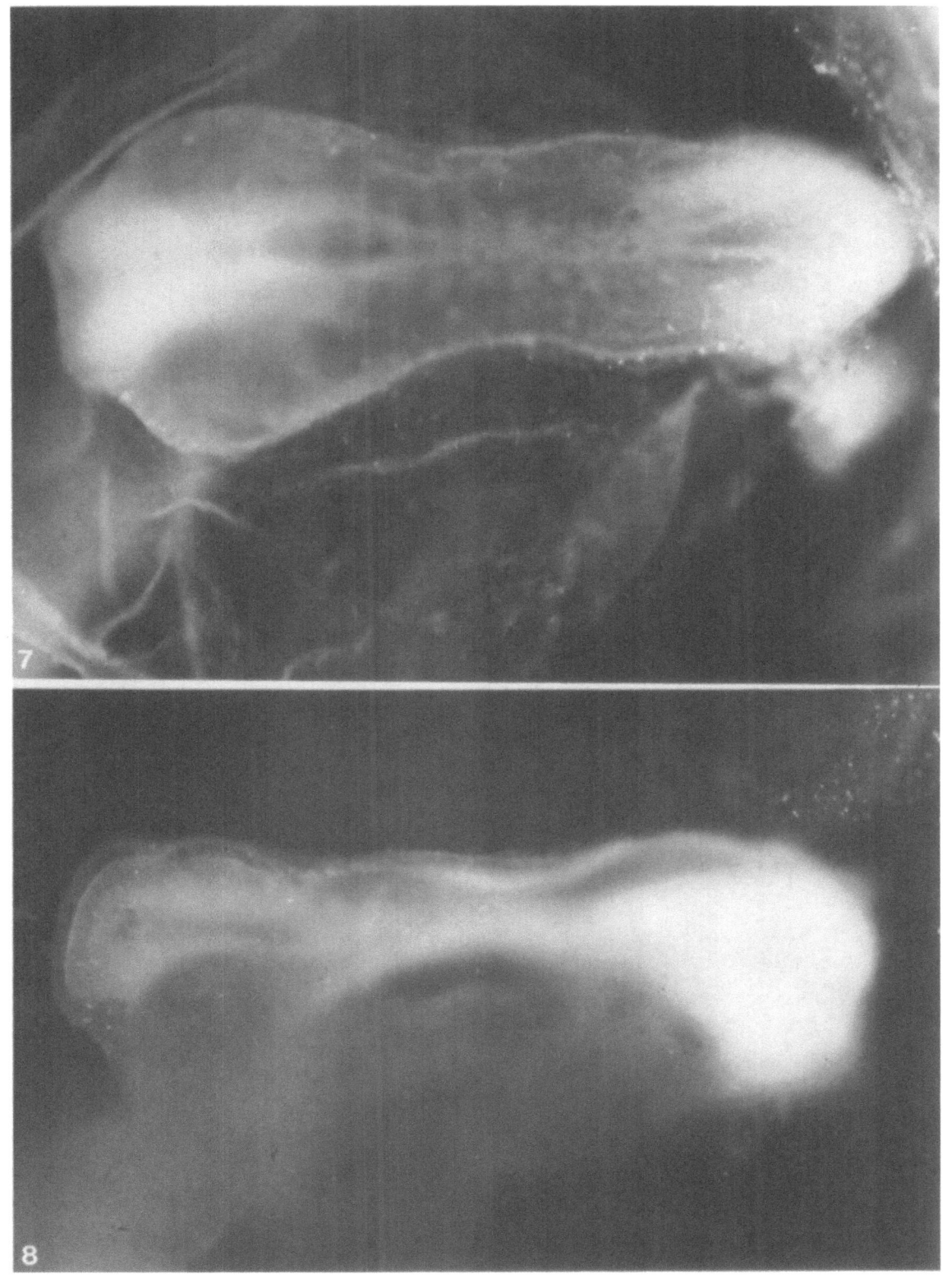

Figures 7 and 8. Embryo 2 mm long, ca. 24 days old, with five paired somites and a just closing neural tube. Stage 6-2. Unfixed specimen: **(7)** Dorsal view. **(8)** Lateral view.

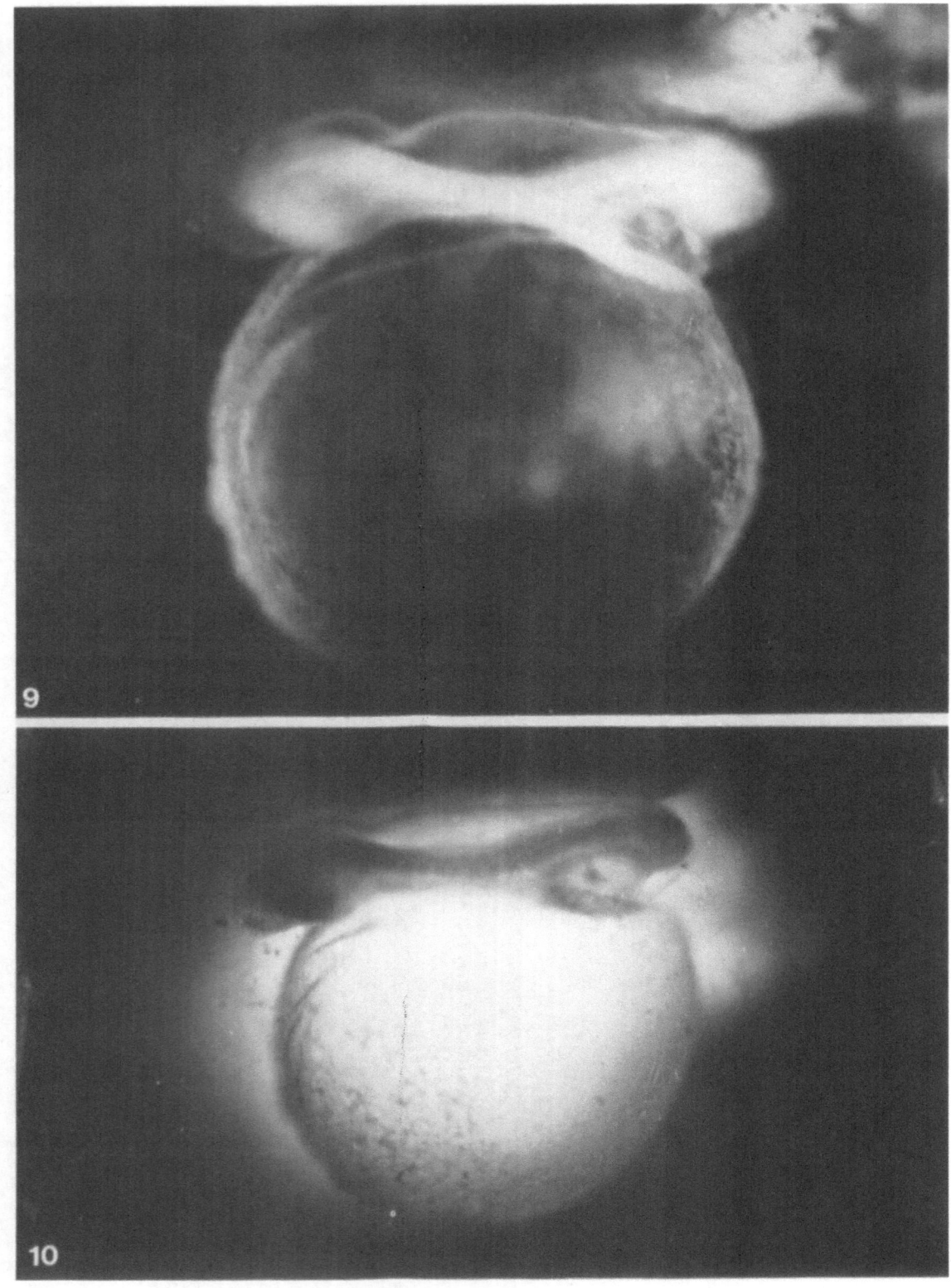

Figures 9 and 10. Embryo 2.5 mm long, ca. 25 days old, with nine paired somites. Stage 6-2. Embryo enclosed within the amniotic sac. A huge yolk sac is attached to the ventral side of the embryo. Unfixed specimen.

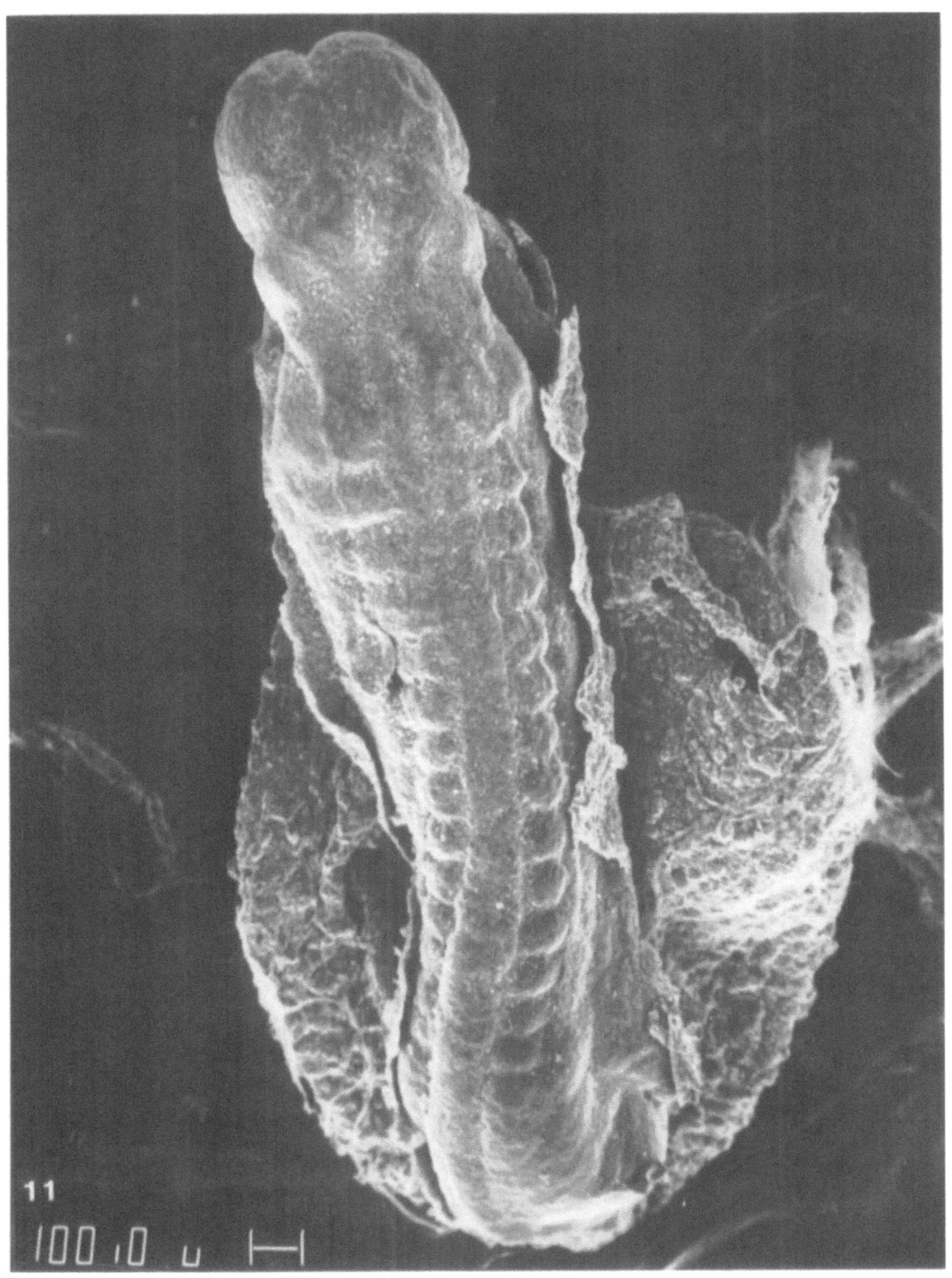

Figure 11. Embryo 3 mm long, ca. 26 days old, with 13 paired somites. Stage 6-2 Amnion dissected. Yolk sac attached to the ventral side of the embryo. Dorsal view. SEM.

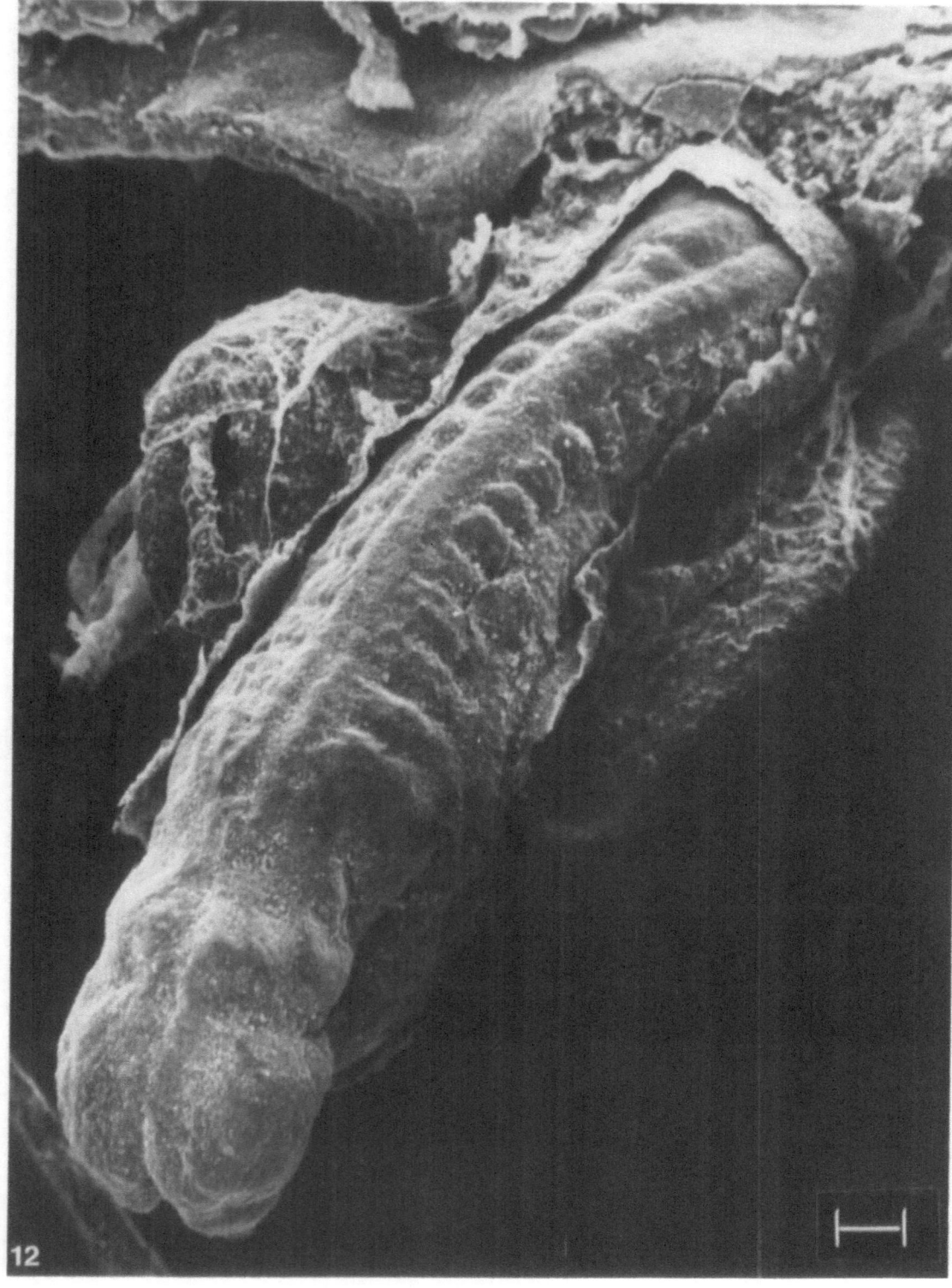

Figure 12. Embryo 3 mm long, 26 days old, with 13 paired somites attached to the chorion. Stage 6-2. SEM.

Stage 6: the early somite stage

Stage 6 (Figs. 13c–15c and 7–16) is characterized by the presence of somites and an open or closing neural tube. The somites are mesodermal vesicles lateral to the neural groove or tube. No limb buds are present. During stage 6, the shape of the embryo is more or less cylindrical at younger stages and becomes C-shaped at older stages. The relatively huge spherical fluid-filled endodermal yolk sac is attached to the ventral side of the embryo. The period in which the neural tube is closing is regarded as the late notogenesis (neurulation).

Substages according to external characteristics:

6-1 Early somite embryo with a completely open neural groove; 1–7 somites. Age 20–21 days, embryonic length 1.5–2.0 mm.

6-2 Early somite embryo with a closing neural tube open anteriorly and posteriorly (first fusion of neural folds occurs at the level of somite 4); 7–20 somites. The first one or two pharyngeal arches are present; otic placodes or pits are evident. Age 21–26 days, embryonic length 1.5–4.0 mm.

6-3 Early somite embryo with anterior or both neuropores closed; 21–30 somites. Two or three pharyngeal arches are present; otic pits are evident. Condensation of the lateral mesoderm precedes formation of the arm buds. The otocysts are closing. Age 26–30 days, embryonic length 3–5 mm.

Stage 7: the stage of limb development

Stage 7 is characterized by the presence of limb primordia and a completely closed neural tube. Somites are distinct in younger substages of this group. (Substages 7–1, 7–2, 7–3, 7–4, and 7–5 are known as late somite stages.)

Substages:

7-1 Embryos with distinct arm buds. Leg buds are inconspicuous; 31–33 somites; otocysts are closed. Three, four, or five distinct pharyngeal arches and the cervical sinus are evident. Age 28–32 days, embryonic length 4–6 mm (Fig. 16c).

7-2 Embryos with arm buds and leg buds; 34–40 somites. Lens pits and olfactory placodes are appearing. Age 31–35 days, embryonic length 5–8 mm (Figs. 17c–19c and 11–17).

7-3 Embryos with bisegmented arm buds (hand segment and a common shoulder–arm segment) and unsegmented leg buds. All somites (42–44 pairs) are formed. Lens vesicles are closed and detached from the surface ectoderm. Olfactory pits are distinct. Age 35–38 days, embryonic length 7–10 mm (Figs. 20c–22c).

7-4 Embryos with bisegmented arm and leg buds. Deep olfactory pits are surrounded by nasal ridges. Segmentation of dorsal mesoderm is still distinct. Age 37–42 days, embryonic length 8–12 mm (Figs. 23C–25C and 12–18).

7-5 Embryos with finger rays on the hand plates; foot plates; deep olfactory pits; and distinct nasal ridges separated by distinct nasolacrimal sulci (pl. from sulcus) from maxillary primordia. Mammary ridges appear. Age 42–44 days, embryonic length 10–14 mm (Figs. 26c and 27c).

7-6 Embryos with finger tubercles separated by interdigital notches on the hand plates; foot plates with toe rays. Segmentation of dorsal mesoderm has completely disappeared. Auricular hillocks are fused to a pinna. Primary palate is closed. Age 44–51 days, embryonic length 13–21 mm (Fig. 28c).

7-7 Embryos with digits and toe tubercles. Age 51–53 days, embryonic length 19–24 mm.

Stage 8: the late embryonic stage

Embryos at stage 8 are characterized by differentiated limbs with all segments including fingers and toes, and open or fusing eyelids.

Substages:

8-1 Embryos with differentiated limbs and open eyelids. The genital membrane is ruptured; the anal membrane is preserved. Age 52–56 days, embryonic length 22–28 mm (figs. 29c, 30c, and 34c).

8-2 Embryos with fusing eyelids. Genital and anal membranes are perforated. External nares are closed by epithelial pluge. Age 56–60 days, embryonic length 27–35 mm (Figs. 31c and 35c; 19–22).

Stage 9: the fetal stage

Stage 9 covers the fetal period between conceptional weeks 9 and 26. The fetus is characterized by

14

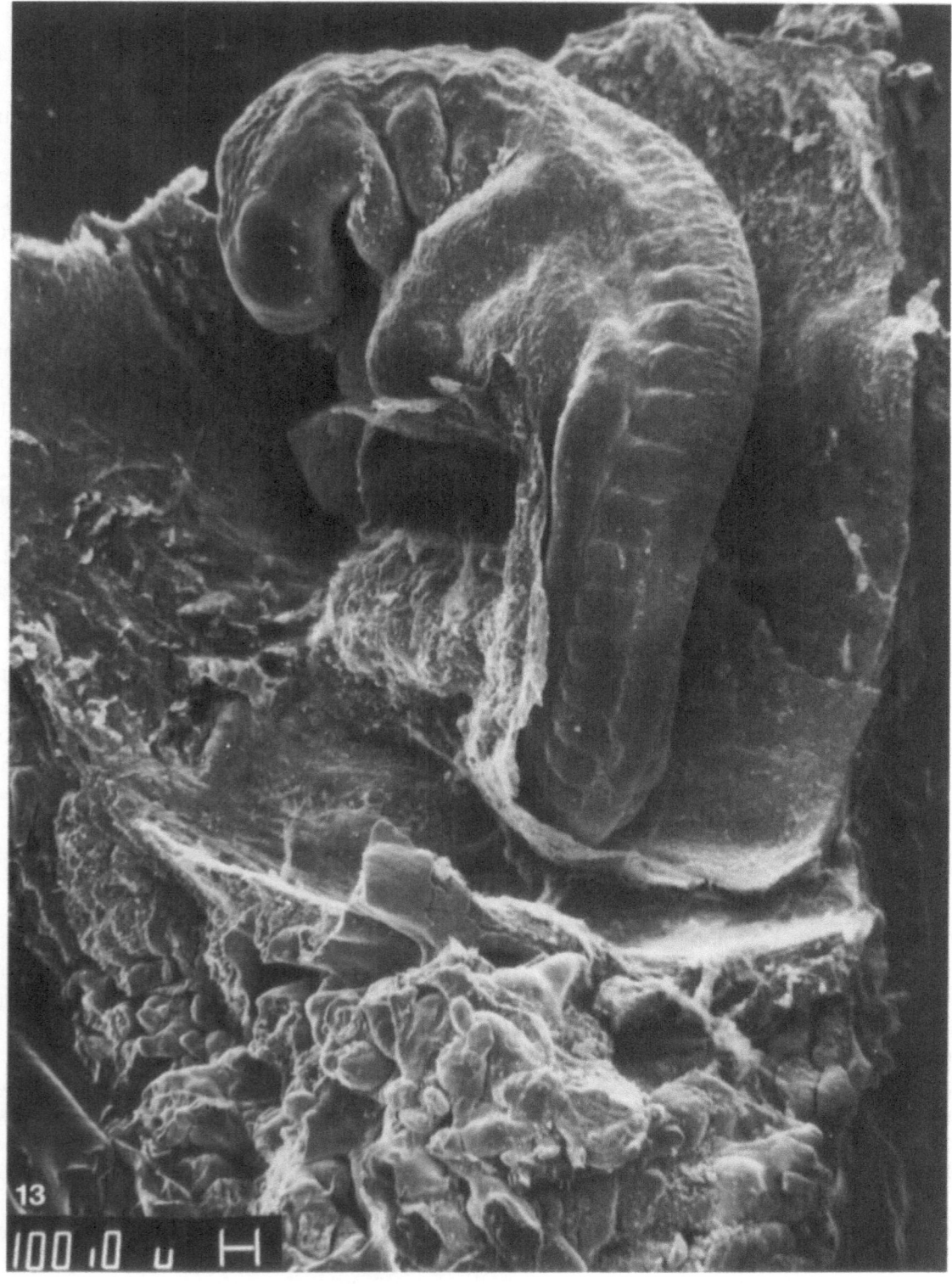

Figure 13. Embryo 4 mm long, ca. 30 days old, attached to the chorion. Stage 6-3. Amnion dissected. No limb buds, three pharyngeal arches, 22 paired somites. Left dorsal view. SEM.

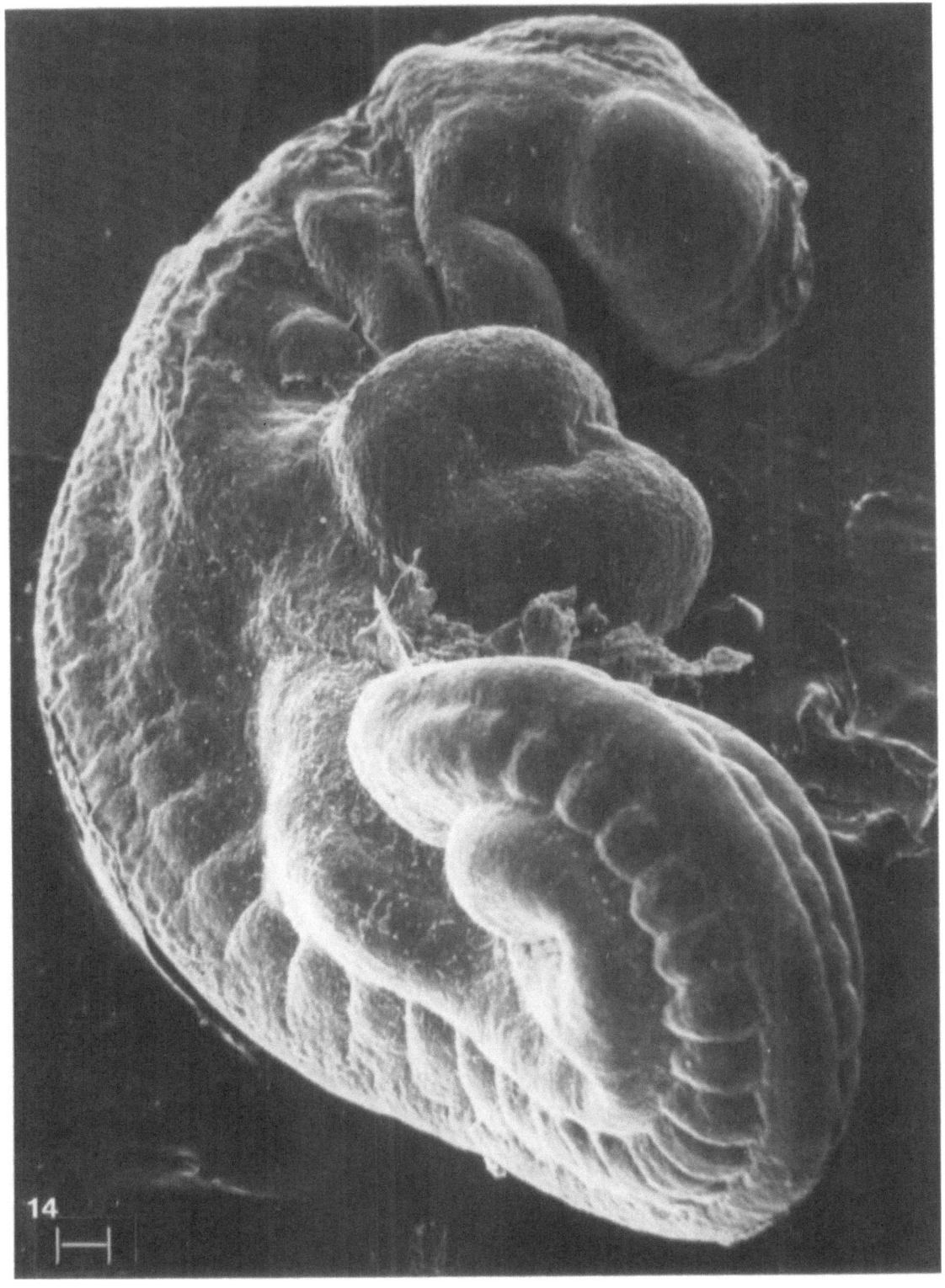

Figure 14. Embryo 4 mm long, 30 dyas old, with 22 paired somites. Stage 6-3. No limb buds, three pharyngeal arches. SEM.

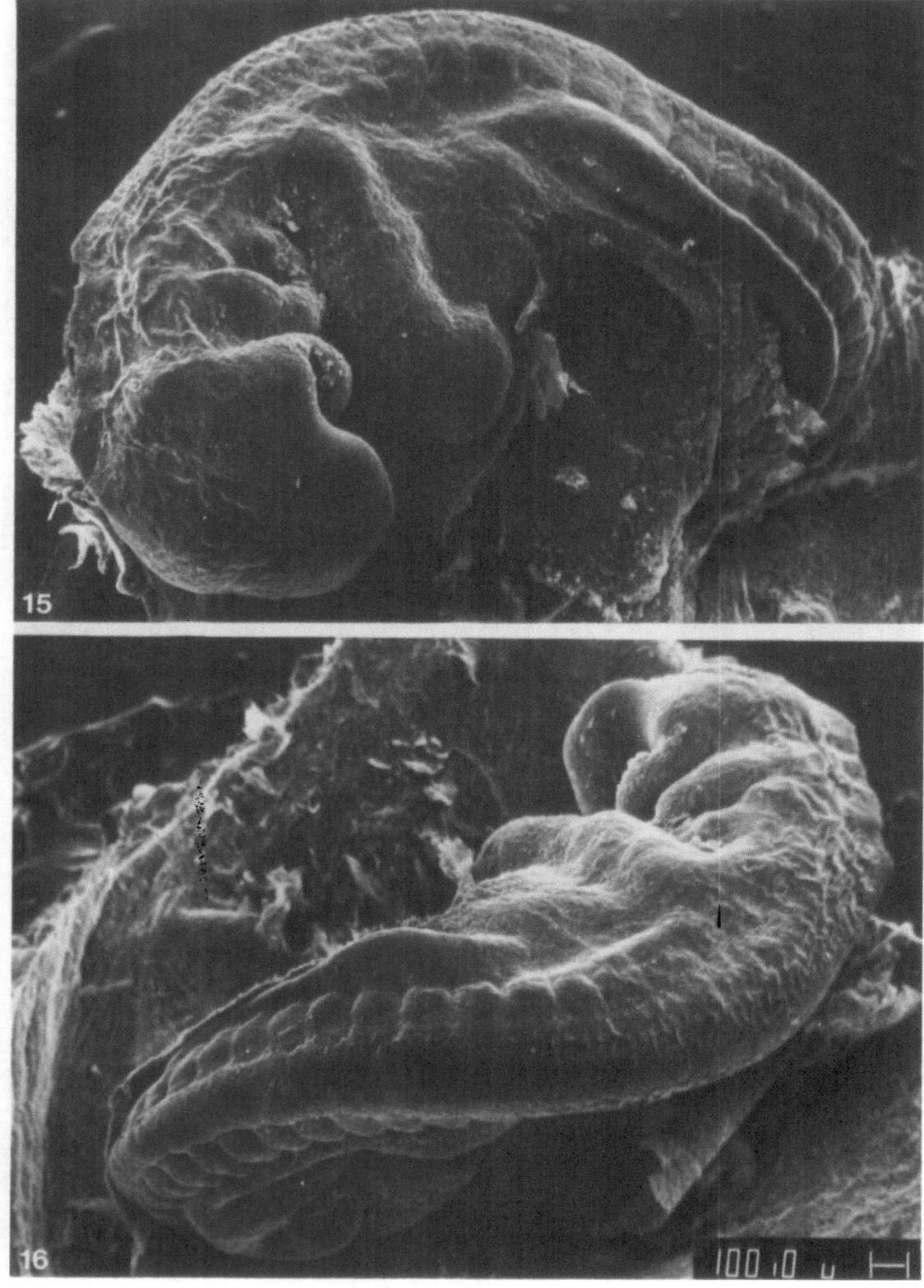

Figures 15 and 16. (15) Embryo 4 mm long, 30 days old, with 22 paired somites. Stage 6-3 Distinct condensation of lateral mesoderm. Left lateral view. SEM. **(16)** The same embryo. Dorsal view. SEM.

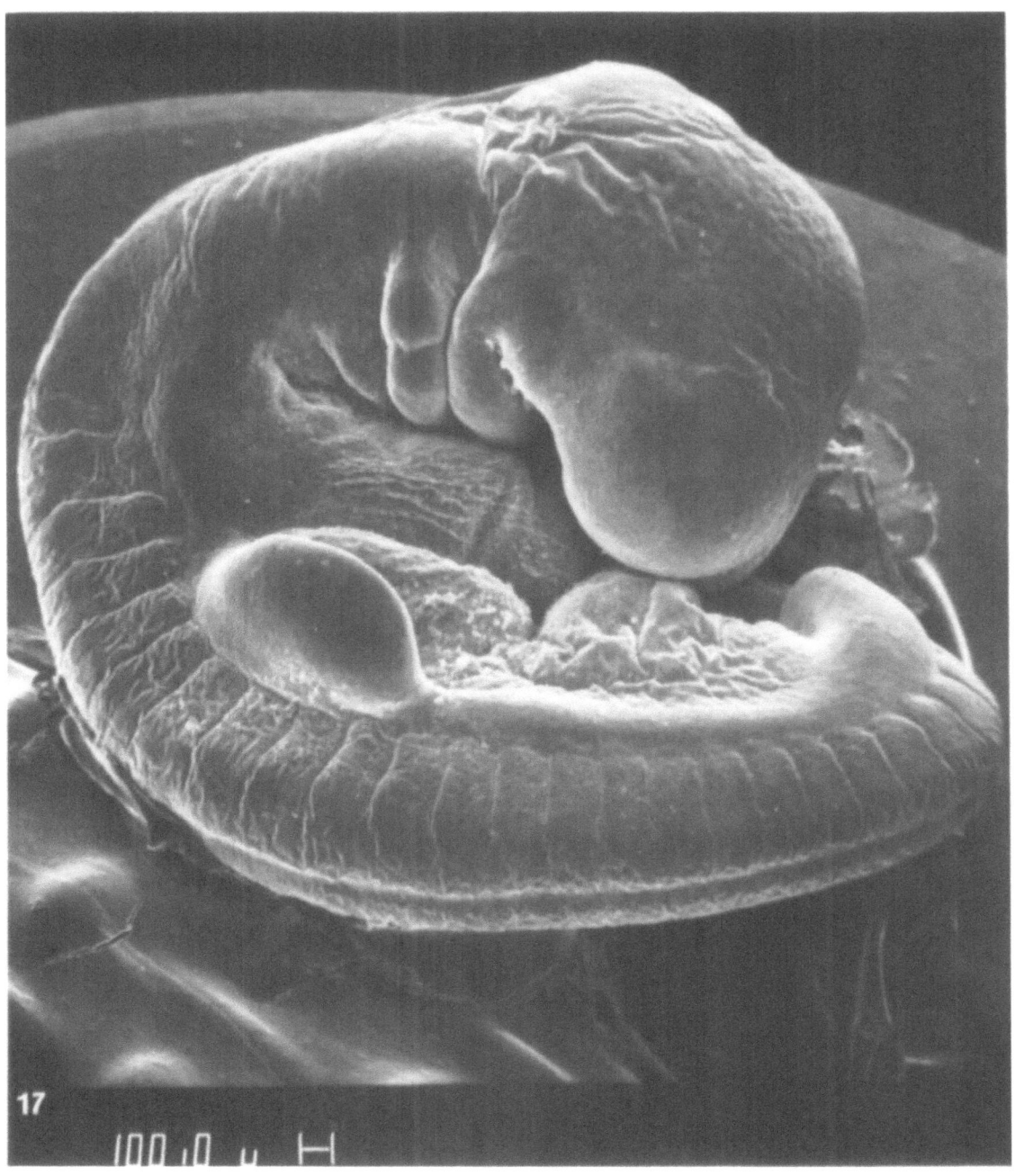

Figure 17. Embryo 5 mm long, 34 days old, with 34 paired somites. Stage 7-2. Buds of proximal and distal extremities: four pharyngeal arches. SEM.

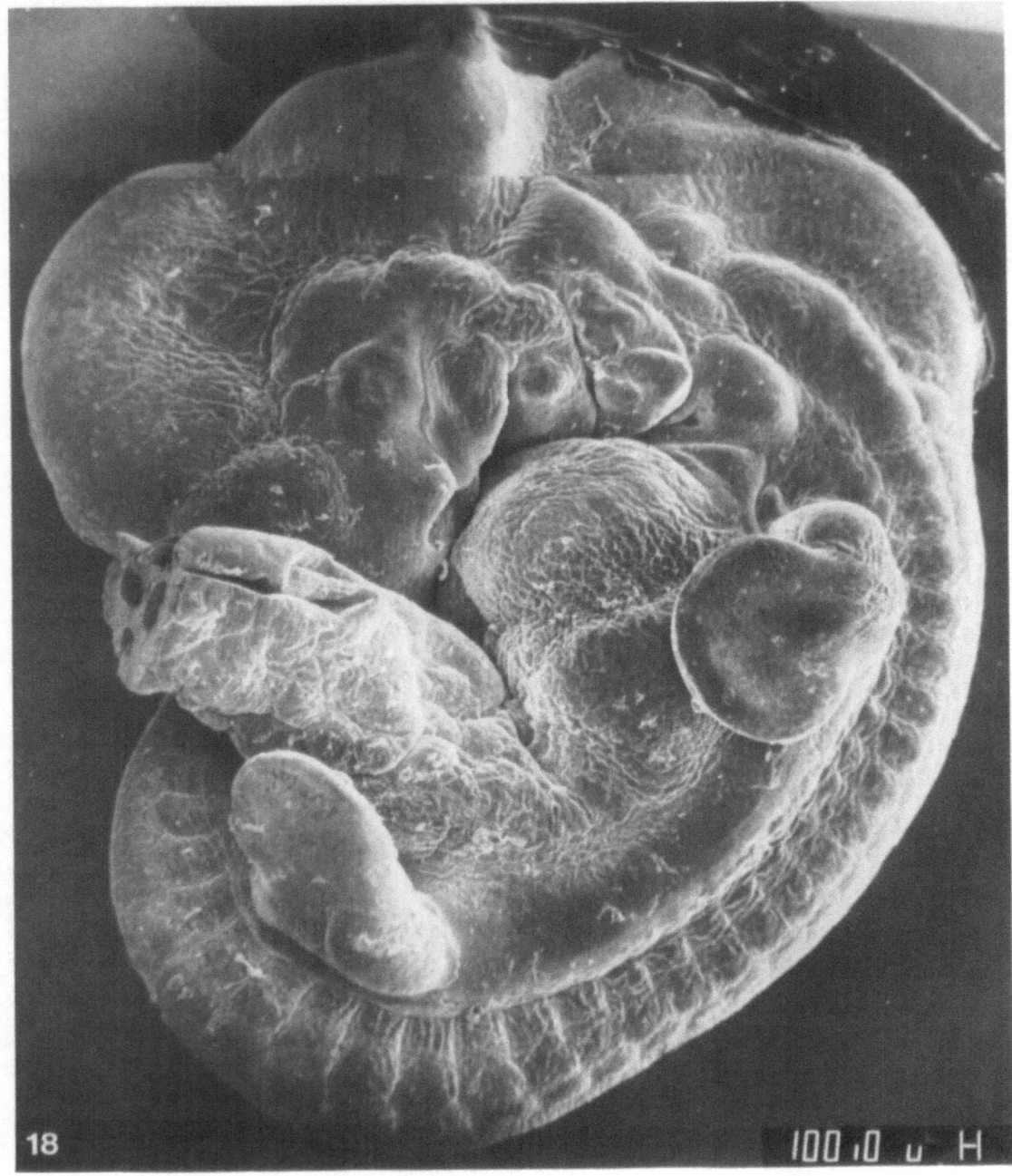

Figure 18. Embryo 10 mm long and 40 days old. Stage 7-4. Bisegmented primordia of anterior and posterior limbs. SEM.

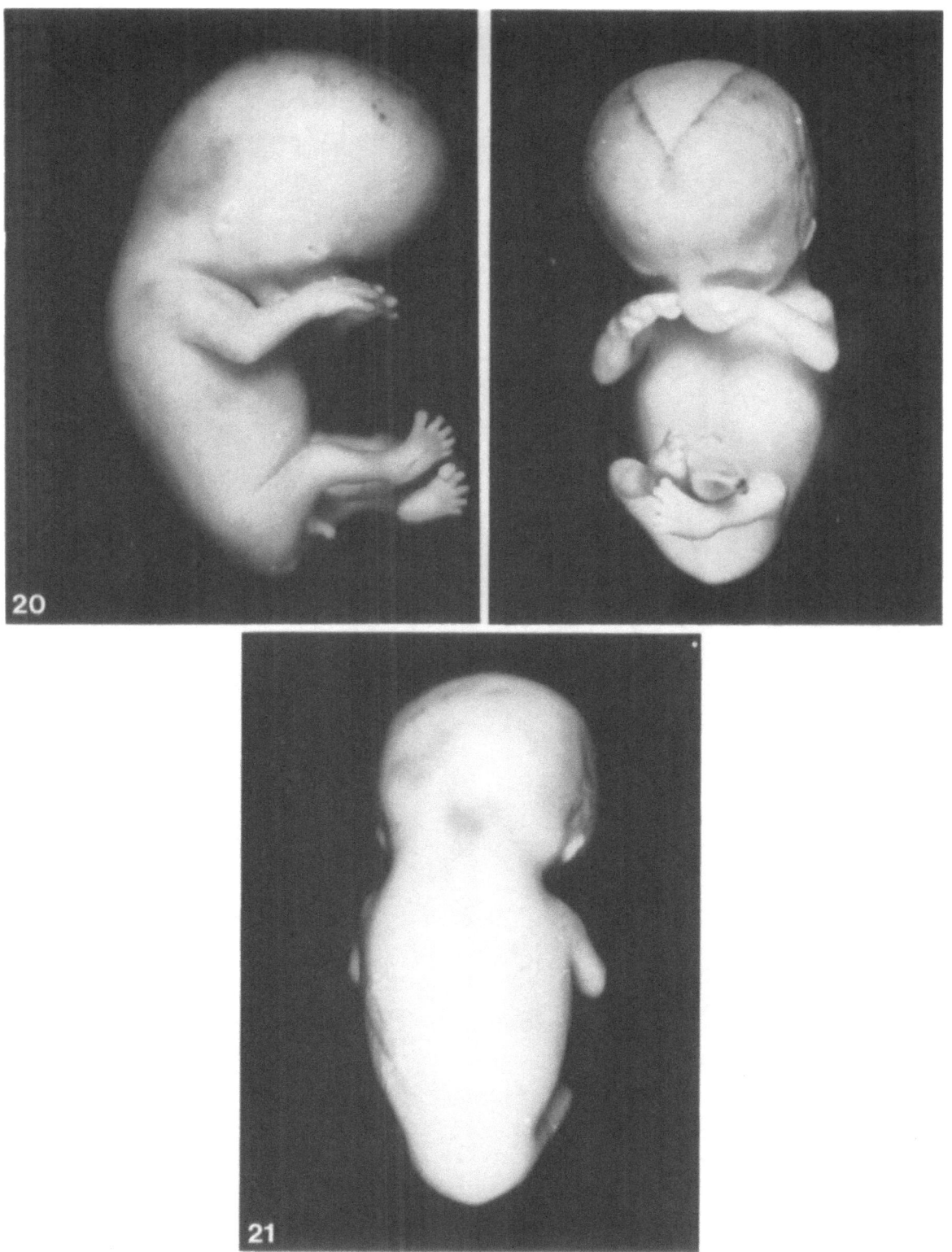

Figures 19–21. Embryo 32 mm long and ca. 58 days old. Stage 8-2. Fully differentiated limbs; fusing eyelids. Unfixed specimen.

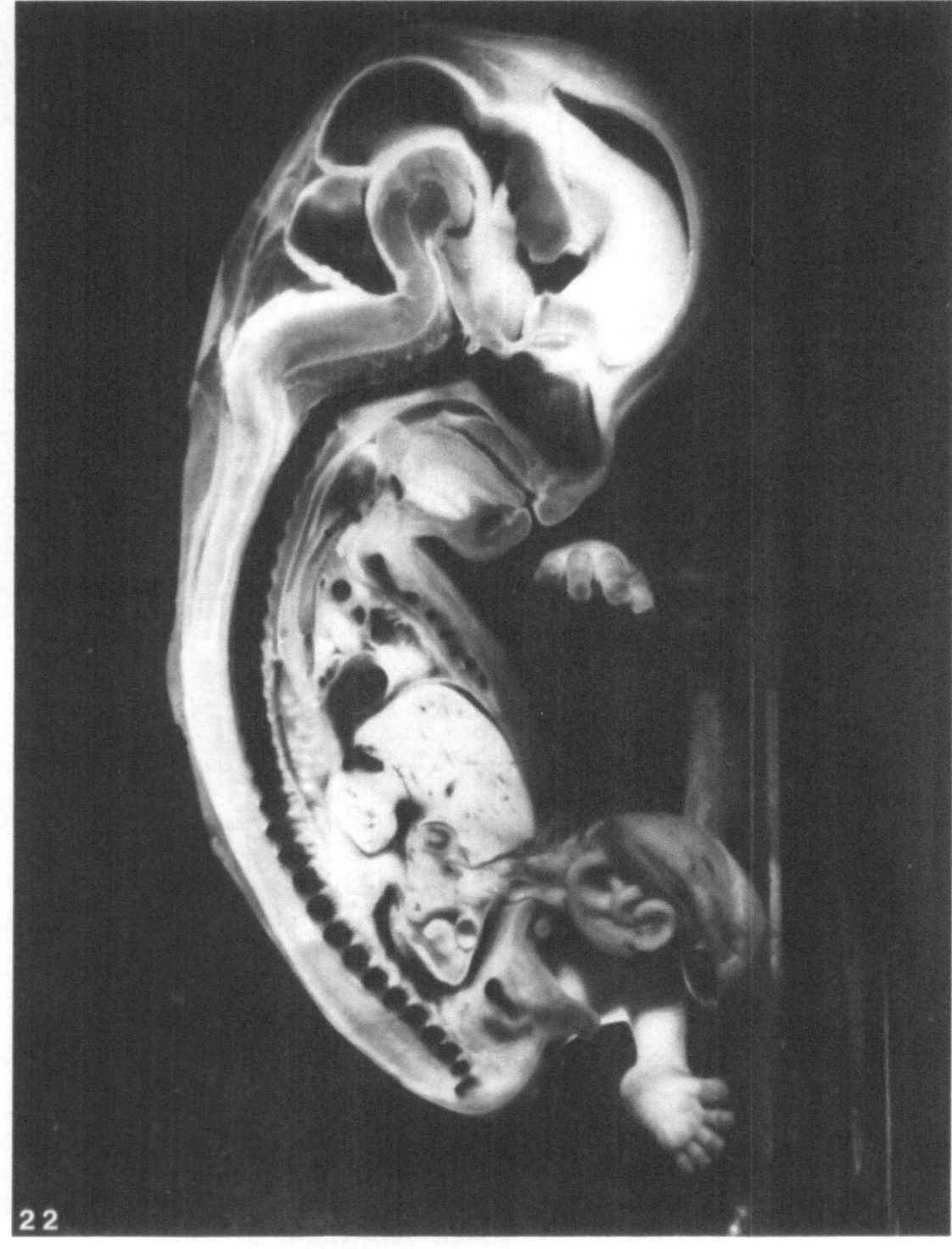

Figure 22. Sagittal dissection of an embryo at stage 8-2, illustrating localization of various organs: brain, hypophysis, medullary tube, cartilaginous cranium, vertebral column, nasal and oral cavities, lung, heart, diaphragm, liver, and gut evaginated partially into the umbilical cord – physiologic hernia – external genitalia.

fused eyelids. At an early fetal period (conceptional weeks 9–12), male differentiation of external genitalia occurs. The weight of the fetus augments from 2 to 50 g. The crown–rump length increases from 30 to ca. 90 mm (figs. 32c, 33c, and 36c).

During the midfetal period (conceptional weeks 13–18), the fetal weight increases from 50 to ca. 300 g. During the late fetal period (conceptional weeks 19–26), fetal weight reaches 1000 g. The crown–rump length reaches ca. 240 mm, the crown–heel length 350 mm.

Stage 10: the perinatal period
The eyelids of the fetus reopen and the total body weight exceeds 1000 g. The crown–rump length is more than 240 mm; the crown–heel length exceeds 350 mm. If the fetus is delivered between 28 and 32 gestational weeks, the term 'immature' newborn is applied; between gestational weeks 33 and 35, the term 'premature' newborn is appropriate; between gestational weeks 36 and 42, the term 'full term' newborn is suitable; if the gestational period exceeds week 42 (more than 40 conceptional weeks), the newborn is called 'postmature.'

Table 1. Timetable and staging of human prenatal development

Periods	Age[a] (days)	Length (mm)	External characteristics	Jirásek's stage	Streeter's horizons (roman) or Carnegie stages (arabic)
Blastogenesis	0–2	0.2	Unicellular (fertilized oocyte)	1	1
	2–4	0.2	Blastomeric (16–20 blastomeres, morula)	2	2
	4–6	0.4	Blastodermic (blastocyst)	3	3
			Bilaminar embryo stage (round embryonic disk)		4
	6–15	0.1	Bilaminar plate	4–1	
			Primary yolk sac	4–2	5
		0.2–0.4	Secondary yolk sac	4–3	6a
			Trilaminar embryo stage (pear-shaped embryonic disk)		6b
	15–17	0.4–1.0	With primitive streak	5–1	
	17–20	1.0–2.0	With a notochordal process	5–2	7–8
Organogenesis			Early somite stage (shoe-sole-shaped embryo)		IX
	20–21	1.5–2.0	Completely open neural groove	6–1	X
			Neural tube closing		
	21–26	1.5–4.0	Both ends open	6–2	XI
	26–30	3–5	One or both neuropores closed	6–3	XII
			Stage of limb development (C-shaped embryo)		
	28–32	4–6	Bud of proximal extremity	7–1	
	3.1–35	5–8	Buds of proximal and distal extremities	7–2	XIII
	35–38	7–10	Proximal extremities: two segments	7–3	XIV–XV
	37–42	8–12	Proximal and distal extremities: two segments	7–4	XVI–XVII
	42–44	10–14	Digital rays, foot plates	7–5	XVIII
	44–51	13–21	Digital tubercles	7–6	XIX
	51–53	19–24	Digits, toe tubercles	7–7	XX
			Late embryonic stage (embryo with differentiated extremities including finger and toes)		XXI
	52–56	22–25	Eyes open	8–1	XXII
	56–60	27–35	Fusing eyelids	8–2	XXIII
Fetal period	60–182+	31–200	Fetus with fused eyelids	9	–
Perinatal period	170–266+	201–350	Third trimester fetus (newborn with open eyes)	10	–

[a] Age from conception.

2. PLACENTA AND PLACENTAL MEMBRANES

The trophoblast of the blastocyst, the trophoblastic shell, and the chorion (Figs. 37c–39c and 23–24)

The main component of the placenta is the trophoblast. External cells of the morula give rise to the trophoblast of the blastocyst. During implantation, the trophoblast of the blastocyst proliferates and changes into the trophoblastic shell composed of cytotrophoblast and syncytiotrophoblast. The trophoblastic shell is implanted within the compact (superficial) layer of the endometrium (interstitial implantation). Lacunae within the trophoblastic shell fuse into the primitive intervillous space supplied by maternal blood from the endometrial spiral arteries and drained by endometrial veins. As the primary mesoderm appears on the inner surface of the trophoblastic shell, the primitive chorion becomes constituted. The trophoblastic shell supported by the primary mesoderm transforms into the (early) villous chorion. Coincidently the maternal, uterochorionic (uteroplacental, intervillous) circulation becomes established. Primary, secondary, and tertiary chorionic villi are distinguished. The primary villi are cords of cytotrophoblast covered by an irregular trophoblastic syncytium within the trophoblastic shell. Primary chorionic villi are converted into the secondary villi by penetration of the avascular mesenchyme (primary mesoderm) into cytotrophoblastic columns. The tertiary villi exhibit capillary nets within the mesenchymal stroma. The chorion is composed of the chorionic plate and chorionic villi. The villi branch extensively (Figs. 37c–39c). The villi branching within the intervillous space are known as free (or resorptive); the villi connecting the chorionic plate with the basal plate (originating from the peripheral portion of the trophoblastic shell and the adjacent decidua) are known as the anchoring villi. Although nets of angioblasts are present in the chorionic villi at the end of the third week, the fetal (embryochorionic, fetoplacental) circulation becomes established during the fifth week postconceptionally, that is, ca. 14 days after the establishment of maternal (uterochorionic, intervillous) circulation (Figs. 5–12).

Beginning three or four days after implantation, the functional layer of the endometrium changes into the decidua. Regarding special relationships to the implanted conceptus, the decidua basalis, marginalis, capsularis, and parietalis are distinguished. As the conceptus enlarges, the capsular decidua becomes apposed to the parietal decidua and the uterine cavity disappears.

Although the trophoblast is nourished by the maternal blood from the intervillous spaces, differentiation and growth of the late chorion depend on the embryochorionic circulation. The chorionic villi oriented toward the basal decidua are supplied preferentially by the embryonic blood. Embryochorionic circulation never develops within the chorionic villi oriented against the decidua capsularis. Those villi that are supplied by embryonic blood give rise to the chorion frondosum. Those villi that do not receive embryonic blood degenerate, and their stroma, including endothelial cords, disintegrate. This area of the chorion not supplied by embryonic blood changes into the chorion laeve.

In relation to the vascular architecture, the following nomenclature is applied to the chorion frondosum: The main (or stem) chorionic branches of the umbilical vessels supply chorionic stems. The secondary, tertiary, etc., chorionic branches contain the secondary, tertiary, etc., branches of the umbilical vessels located within the mesenchymal stroma in a paraxial localization. The 'terminal villi' are projections on the surface of the various chorionic branches (including terminal chorionic branches) that contain only capillary loops under-

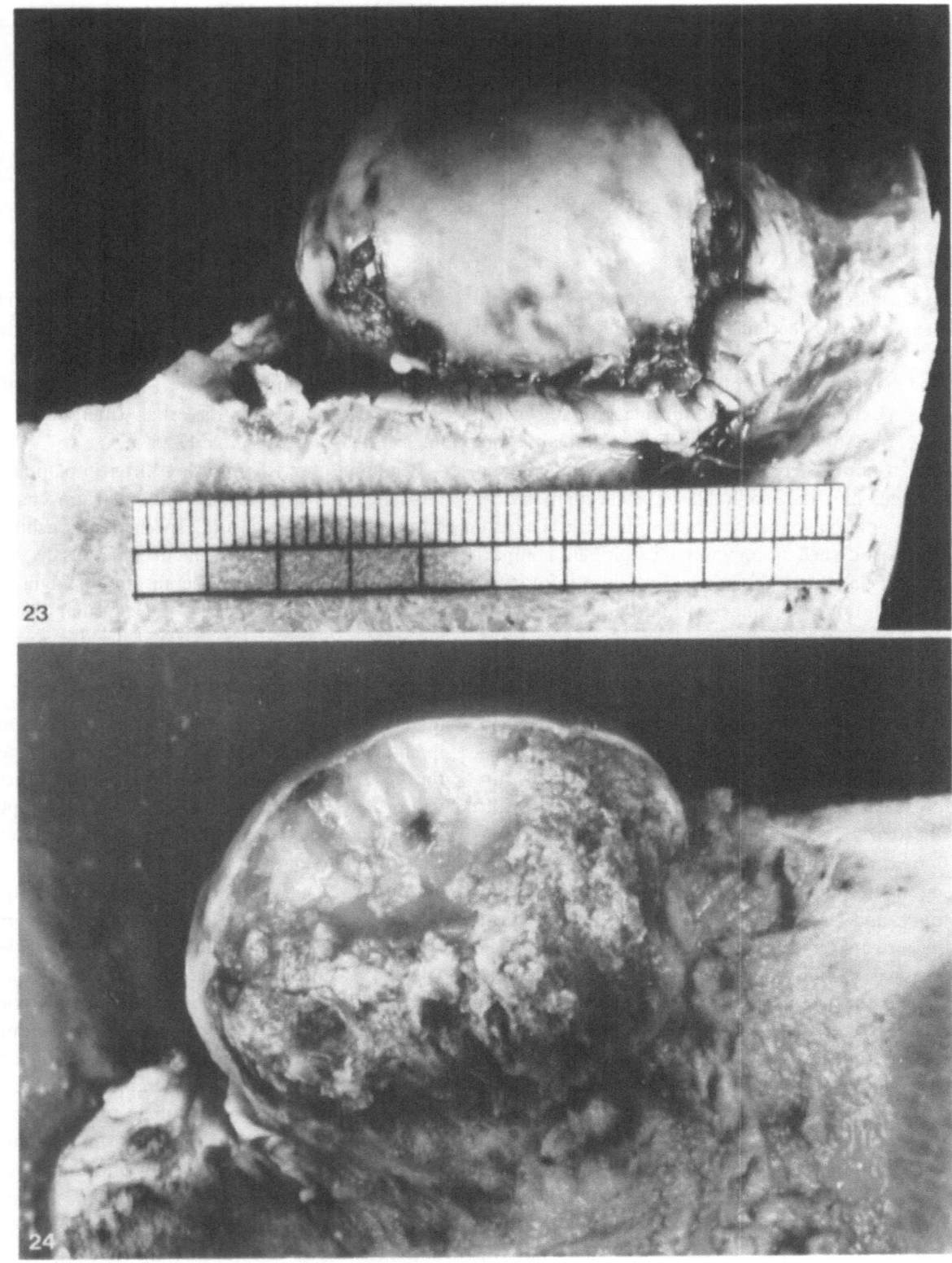

Figures 23 and 24. (23) Conceptus implanted within dissected uterine wall. Stage 8-5. (24) Implanted conceptus. Stage 8-5. The decidua was removed; chorion laeve and frondosum are evident.

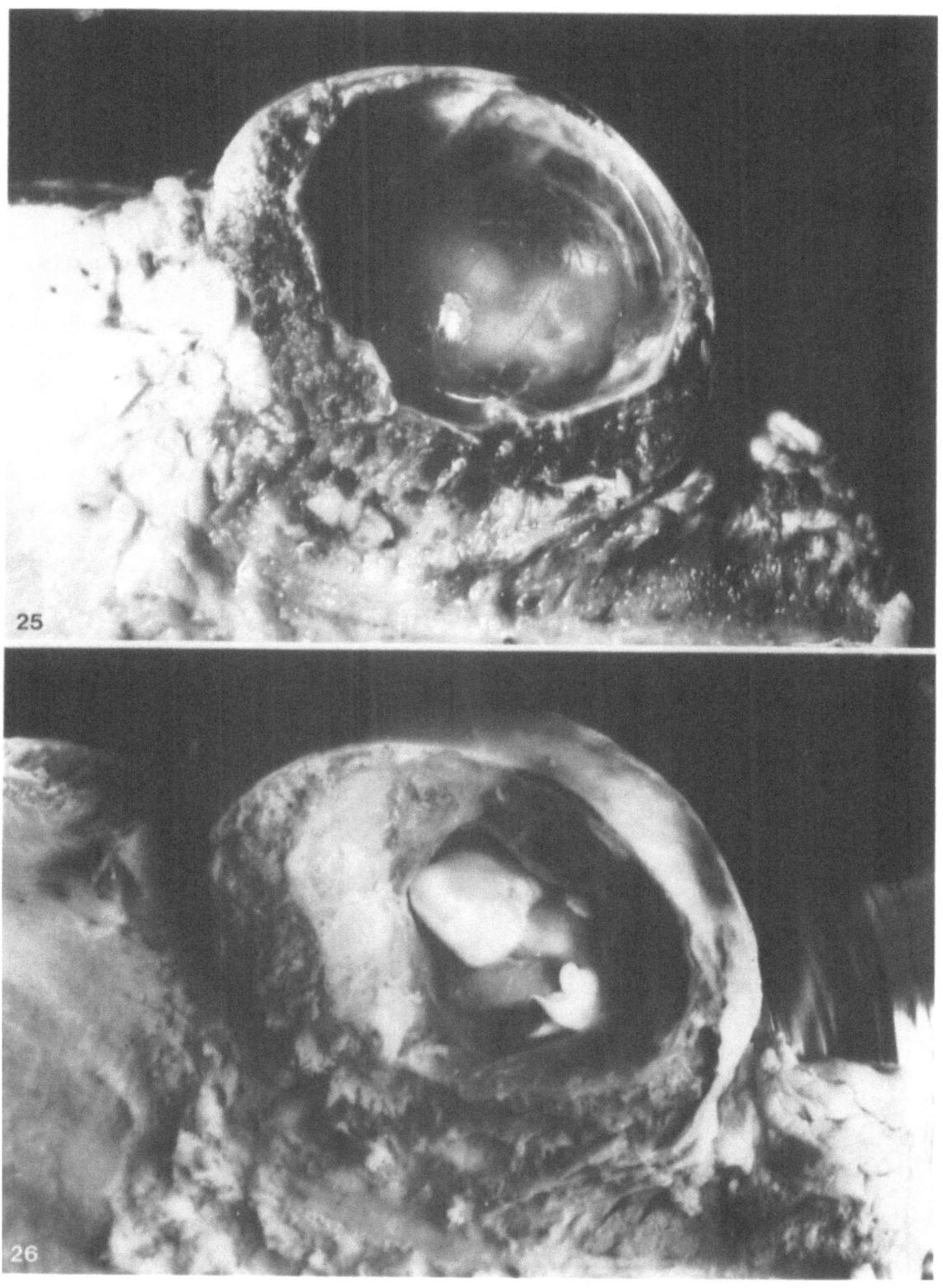

Figures 25 and 26. The plancenta. **(25)** Implanted conceptus with dissected chorionic and amniotic sacs exposed. Stage 8-5 **(26)** Implanted conceptus with dissected decidua and chorionic and amniotic sacs. The embryo present within the amniotic sac is attached to the chorion by the umbilical cord. Stage 8-5.

26

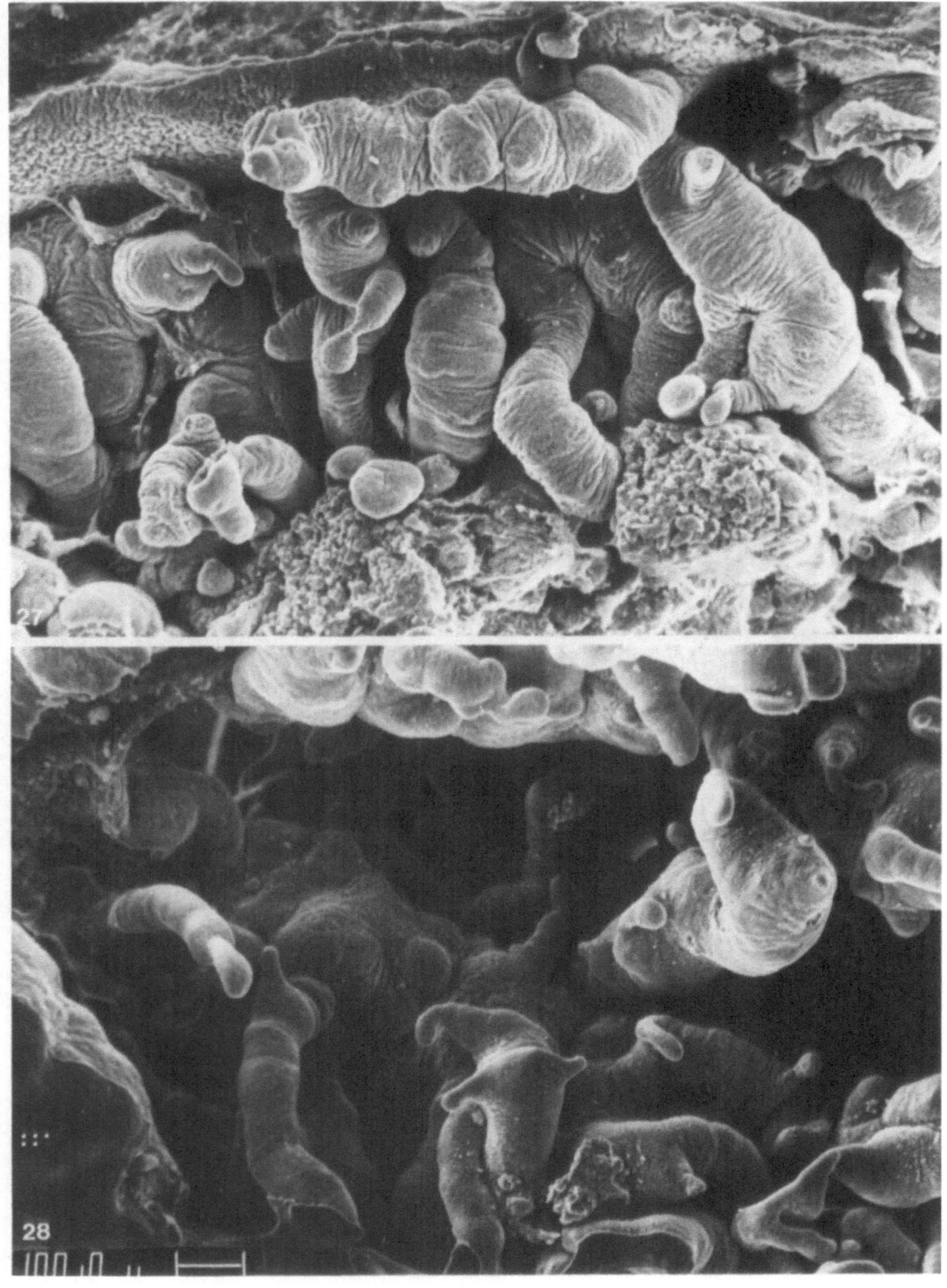

Figures 27 and 28. (27) Chorionic plate, chorionic villi with syncytial buds, and cytotrophoblastic islands near the surface of basal decidua. Stage 7-2. SEM. **(28)** Chorionic villi with syncytial buds. Stage 7-2. SEM.

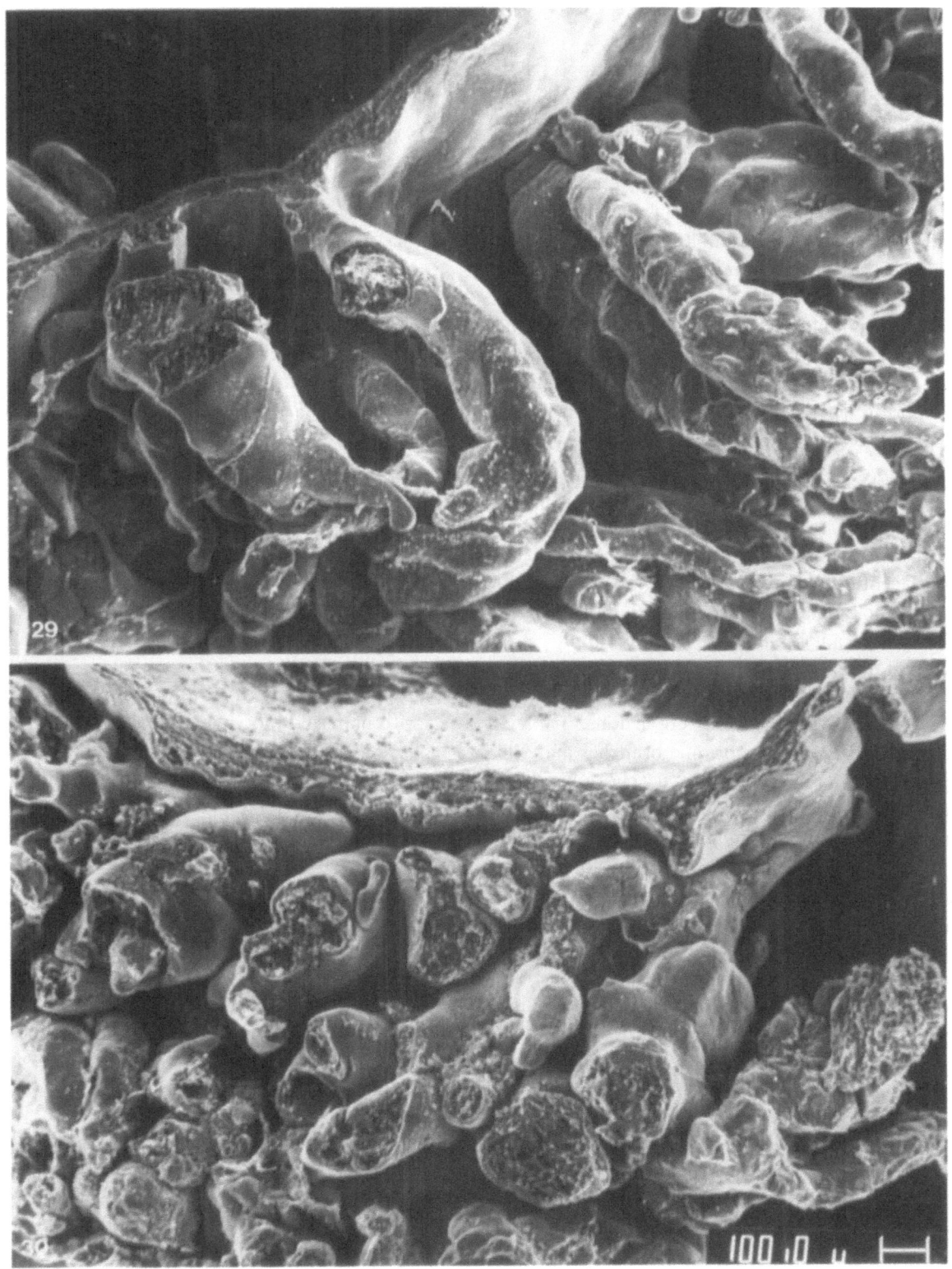

Figures 29 and 30. (29) Chorionic villi extending from the chorionic plate into the intervillous space. Stage 7-2. SEM. **(30)** Chorionic plate and chorionic villi. Stage 7-2. SEM.

neath the trophoblast. Capillary nets apposed to the trophoblastic syncytium are sites of intensive maternofetal and fetomaternal exchange and are known as the syncytiovascular membranes. The chorionic stroma is nourished from fetal blood.

Fetal components of the placenta are derived from the chorion frondosum and are covered from inside by the placental portion of the amnion. The maternal portion of the placenta is contributed to by the trophoblastic basal plate fused with the decidua basalis. The placental membranes originate from the decidua (fused parietal and capsular) and the chorion laeve, which is covered from inside by the extraplacental portion of amnion.

The placenta

The placenta is a discoid organ composed of a fetal portion and a maternal portion. The uterine surface of the placenta is lobulated. There are about 15–30 placental lobules incompletely separated by decidual septa coming from the basal plate. Each lobule includes two or three cotyledons. The cotyledon is a system of chorionic branches and villi supplied by arteries from a primary chorionic stem. The fetal surface is smooth, covered by the amnion. The fetal portion of the placenta comprises the placental amnion fused with the chorionic plate and the cotyledons. Among chorionic villi and branches, there is the intervillous space containing maternal blood supplied by coiled arteries of the endometrium. The endometrial arteries penetrate the basal plate and open underneath each cotyledon more centrally than the uterine veins draining blood from the intervillous space. The basal plate of the placenta, in spite of a mixed fetal and maternal origin, is considered as the maternal part of the placenta. Most of the maternal portion of the placenta is the (maternal) basal decidua. Some cells facing the intervillous space are trophoblastic derivatives, fetal in origin. Trophoblastic derivatives are usually separated from the decidua by a sheet of an amorphous substance, the basal fibrinoid. According to anatomy, the human placenta is considered discoidal, cotyledonic, decidual, and hemochorial. The normal size of a full-term placenta is ca. 150 × 250 × 30 mm and the weight ca. 400–600 g. The weights of the placental active mass

and of the fetus are related. The size of the placenta seems to be related to the minute volume of the fetal heart.

Survey of placental components
1) *Placental (chorionic) plate*
   a) Placental portion of the amnion: (i) amniotic epithelium, and (ii) mesenchymal amniotic propria
   b) Chorionic plate: (i) mesenchymal propria with main vessels, and (ii) trophoblast of the chorionic plate
2) *Placental 'villi'* –anchoring and free (primary chorionic stems; primary, secondary, and tertiary chorionic branches; and terminal villi cotyledons)
   a) Mesenchymal propria of the villi with fetal vessels
   b) Trophoblast of the villi
3) *Placental basal plate* with placental septa
   a) Basal trophoblast
   b) Basal decidua
   Intervillous space: space among the villi is filled with maternal blood (marginal sinus – lateral border of the intervillous space; subchorionic lake – portion of the intervillous space underneath the chorionic plate)
   Fibrinoid (amorphous material containing fibrin deposited in various areas of the placenta)
   a) Subchorial fibrinoid (of Langhans, within the chorionic plate)
   b) Villous fibrinoid (of Rohr, on the placental villi)
   c) Basal fibrinoid (of Nitabuch, within the basal plate)

The placental membranes
The placental membranes, which originate from the extraplacental amnion and chorion laeve fused with the decidua capsularis and parietalis, are composed of the following layers:

1) Amnion
   a) Amniotic epithelium
   b) Mesenchymal amniotic propria
   c) Mesenchymal avascular spongy layer derived from mesenchyme of the chorionic cavity
2) Extraplacental chorion

a) Chorionic mesenchymal propria
b) Extraplacental cytotrophoblast
3) Decidua
a) Capsular
b) Parietal

The umbilical cord (Figs. 35–37)

The early embryo is attached to the chorion by a mesenchymal connecting stalk containing the allantois and primordia of the umbilical vessels: two arteries and two veins. Anteriorly to the body stalk, there is the yolk sac connected with the gut of the embryo by the omphaloenteric duct, which is accompanied by vitelline vessels (two arteries and two veins). The body stalk and the omphaloenteric duct are located within the extraembryonic coelom and are pushed together by the expanding amnion. As the amniotic cavity enlarges, the extraembryonic coelom disappears and the mesenchyme of the expanding amnion fuses with mesenchyme located around the omphaloenteric duct and with mesenchyme of the connecting stalk. At later stages, the right umbilical vein as well as all the vitelline vessels and the omphaloenteric duct disappear. A remnant of the allantoic canal is usually present in the proximal portion of the umbilical cord. Loops of the midgut extend temporarily (at conceptional weeks 8–10) into the proximal portion of the umbilical cord as a physiologic umbilical hernia (Figs. 14 and 15).

At term, the umbilical cord is ca. 50 cm long, 2 cm thick, and covered by a single-layered amniotic epithelium. Three umbilical vessels (two arteries and one vein) are tortuous, located within the jelly of Wharton, a myxomatous connective tissue of mesenchymal origin.

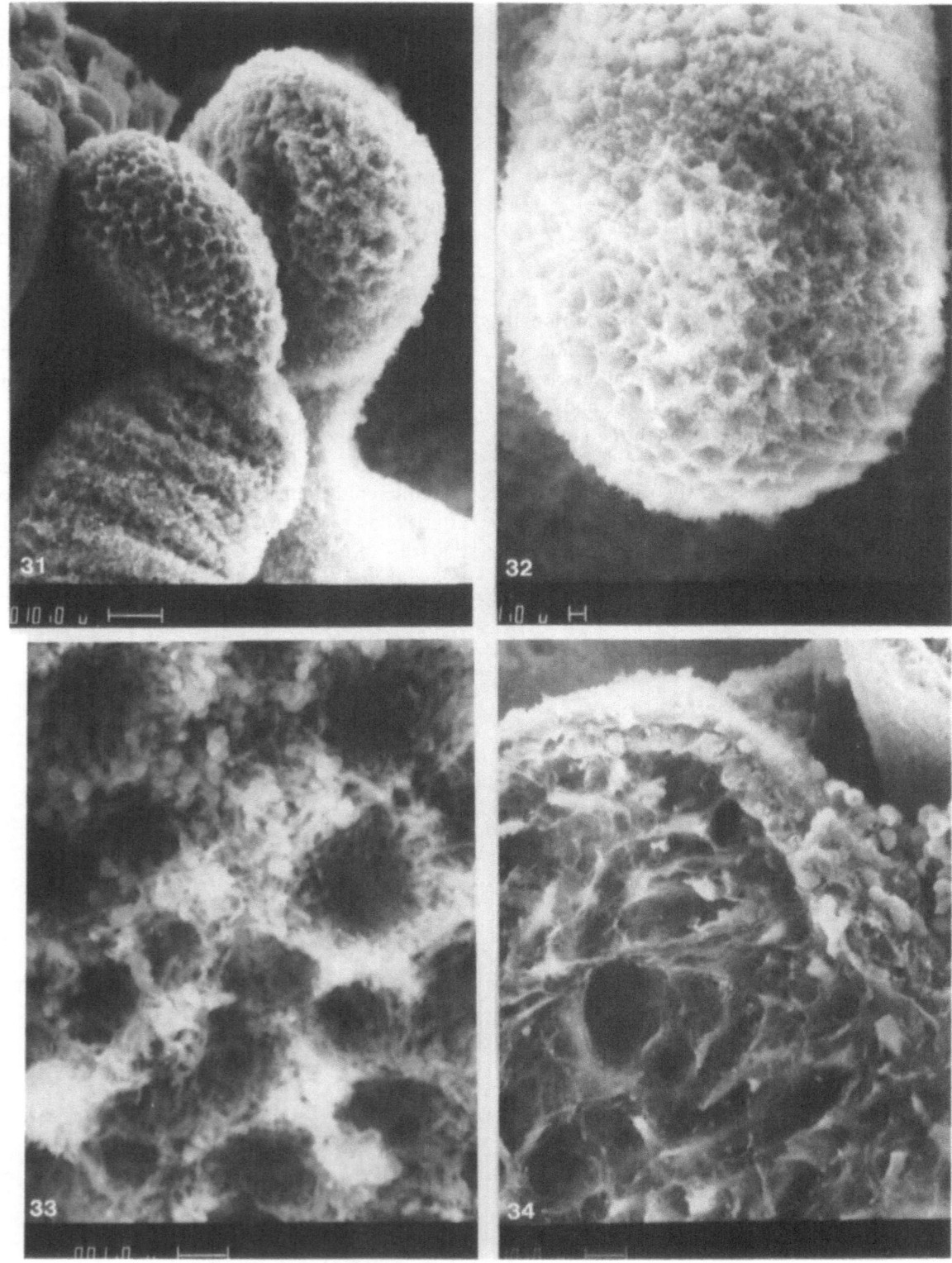

Figures 31–34. (31) Surface of syncytial trophoblastic buds. Stage 7-2. SEM. (32) Pits and cristae on the surface of a trophoblastic bud. Surface structures are formed by microvilli and globular particles. (33) Globular particles ca. 0.2 μ in diameter present on the surface of pits formed by microvilli. (34) Dissected chorionic villus; mesenchymal stroma covered by cytotrophoblast and trophoblastic syncytium. Several maternal red blood cells are in the intervillous space. SEM.

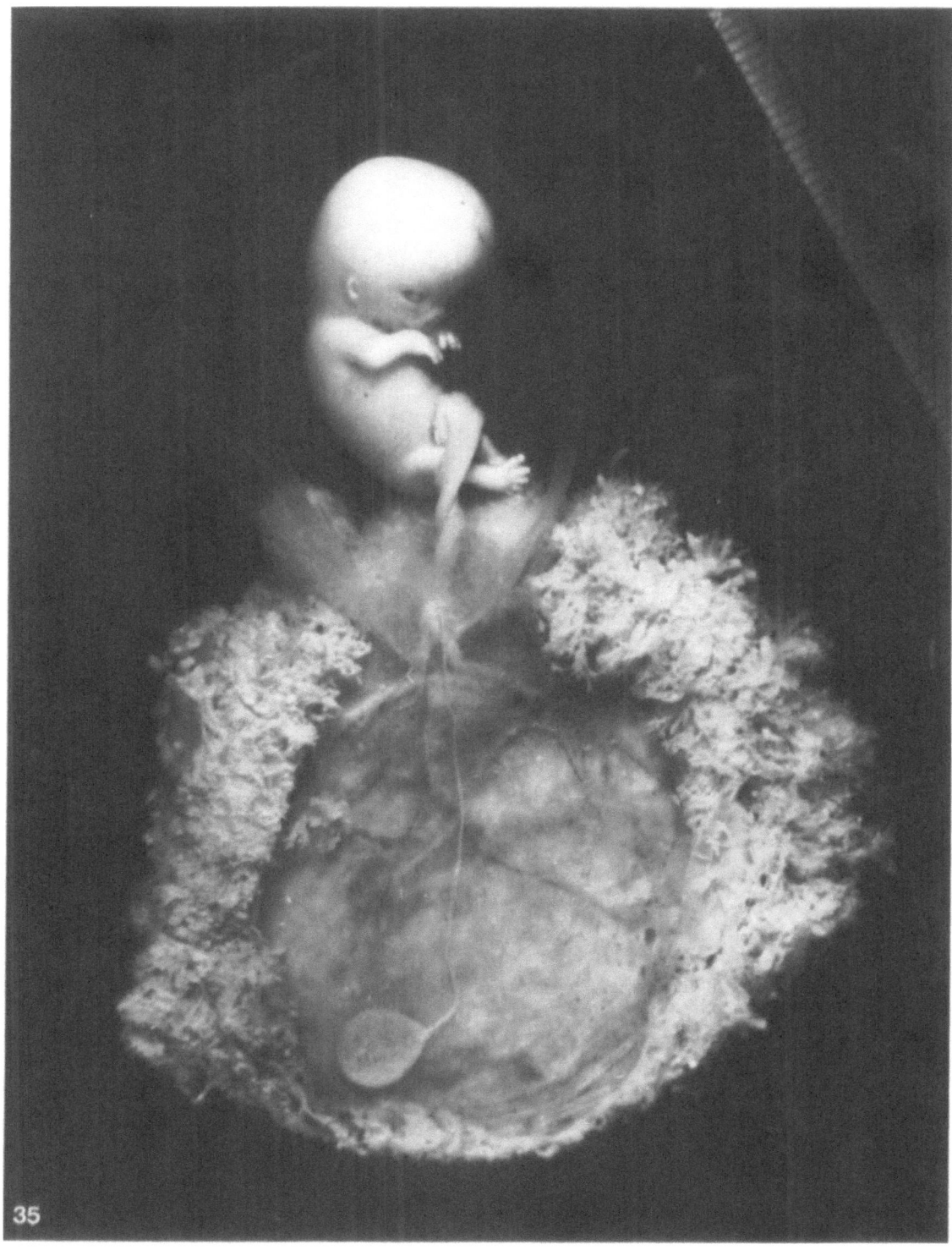

35

Figure 35. Embryo with envelopes. Stage 8-1. The embryo enclosed in a transparent dissected amnion is attached to the chorion by the umbilical cord. A yolk sac with a long vitteline duct is attached to the chorionic end of the umbilical cord.

32

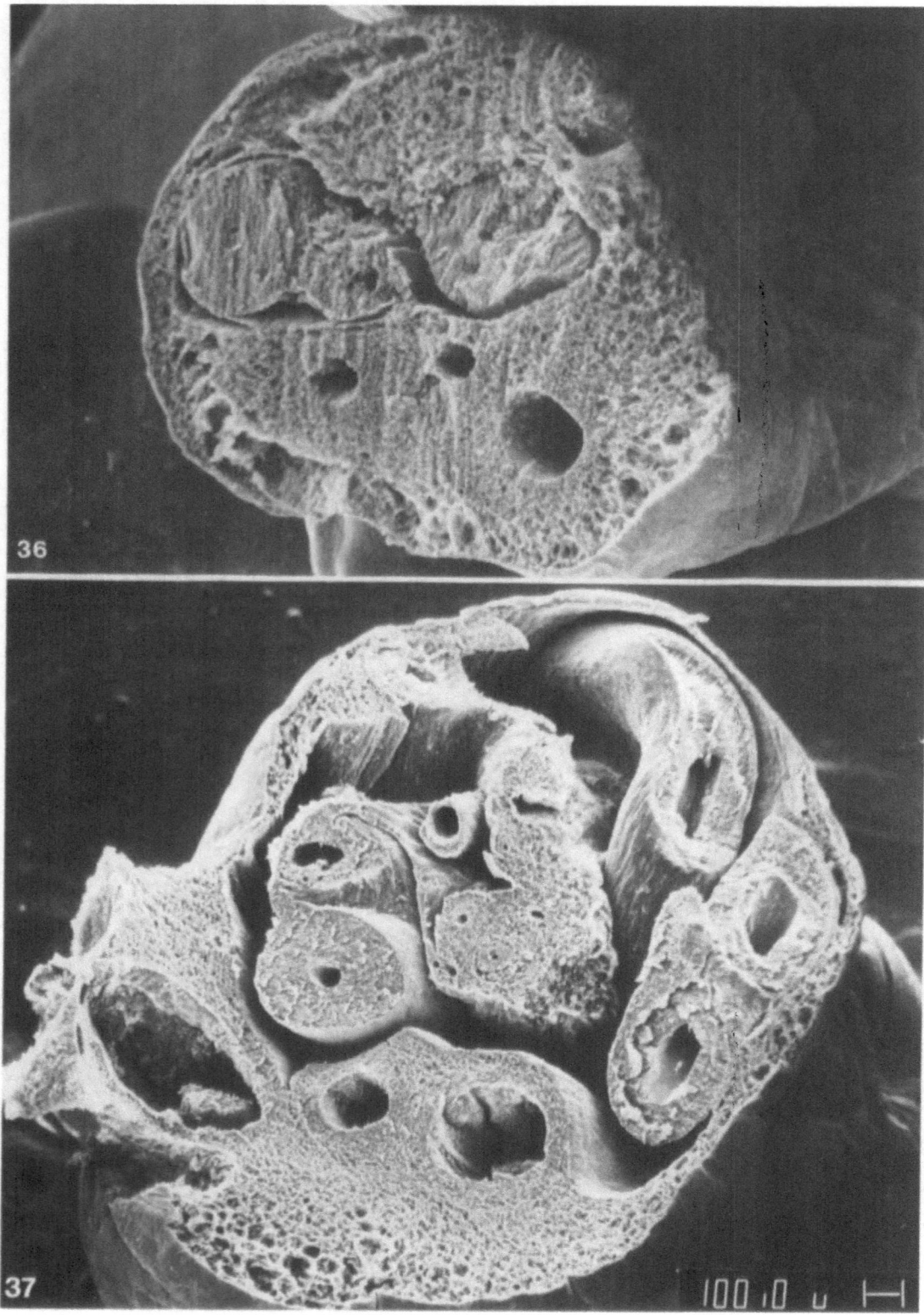

Figures 36 and 37. Proximal portion of the umbilical cord. **(36)** Stage 7-4. Anteriorly the amniotic mesoderm is fused with mesoderm located around the vitelline duct and posteriorly with the mesoderm of the connecting stalk containing two umbilical arteries and the allantois. A primitive intestinal loop is present within the exocoelom. **(37)** Stage 8-1. Two umbilical arteries and the allantois are present posteriorly within the amniotic mesoderm. Physiologically herniated intestinal loops are located within the extraembryonic coelom of the umbilical cord.

3. GERM LAYERS

The concept of three germinal layers – the ectoderm, the endoderm, and the mesoderm – created by descriptive and comparative embryologists during the second half of the nineteenth century greatly facilitated the teaching of embryology and the understanding of similarities in ontogenic and phylogenic development in different species of animals. According to this concept, the fertilized oocyte gives rise to a cellular aggregate of more or less undifferentiated blastomeres (the morula); the blastomeres subsequently arrange in a sphere (the blastula) lined by the primary embryonic epithelium, the blastoderm. The blastoderm differentiates into two primary embryonic layers: the early ectoderm and endoderm. The bilaminar embryonic stage in lower vertebrates is known as the gastrula. The endoderm-lined cavity within the gastrula is the gastrocoele (the primary cavity of the body). The gastrocoele has a single external opening, the blastopore. In mammals, the morula transforms into the blastocyst. The third embryonic layer, which develops between the ectoderm and the endoderm, is the chordomesoderm. In contrast to the ectoderm and the endoderm, the mesoderm never develops into a 'true epithelium' with a basement membrane. The notochord supports the neuroectodermal plate and induces differentiation of the neural tube. Through interactions with neighboring structures, this portion of mesoderm located among the neuroectoderm, the surface ectoderm, and the endoderm differentiates into paired vesicles known as the somites, while mesoderm located between the surface ectoderm and endoderm splits into the parietal layer (the somatopleura) and the visceral layer (the splanchnopleura). Mesodermal cells of somites and pleurae exhibit 'epithelial' differentiation only apically, facing the coelomic cavity. The basal portions of mesodermal cells are attached by long processus to the basement membranes of ectodermal or endodermal structures. The somites and the primitive pleurae represent pools of cells that are able to migrate, filling spaces between germ layers. These migrating cells constitute the mesenchyme.

Embryologists who distinguish between germ layers still capable of further segregation use the terms 'ectoblast', 'endoblast', and 'mesoblast'. From a descriptive viewpoint, the terms 'ectoblast' and 'endoblast' should be reserved for components of embryos at the bilaminar stage. The term 'mesoblast' should be reserved for cells constituting somites and pleurae still able to provide mesenchymal cells.

Most controversies among embryologists evolved from developmental heterochrony of germ layers occurring in higher vertebrates, especially mammals, including man. Experimental results indicate that there is not so much rigidity and specificity in the fate of various components of germ layers. In all experiments, the prospective potency (e.g. differentiation in one way or another) must be clearly distinguished from the prospective significance(e.g., way of differentiation under normal conditions).

Human development is characterized by a precocious development of extraembryonic structures, especially the trophoblast. Differentiation of the blastocystic trophoblast precedes differentiation of embryonic layers. The inner cell mass (embryoblast) represents a pool of cells that are able to differentiate into all embryonic tissues. From this pool, different cells are detached at different periods of embryonic development. First are detached the endodermal cells of the yolk sac, while other endodermal cells and primary germ cells, which temporarily colonize the hindgut, immigrate to the 'primary' endoderm at a later stage. In addition to germ cells, there might be other pools of cells

34

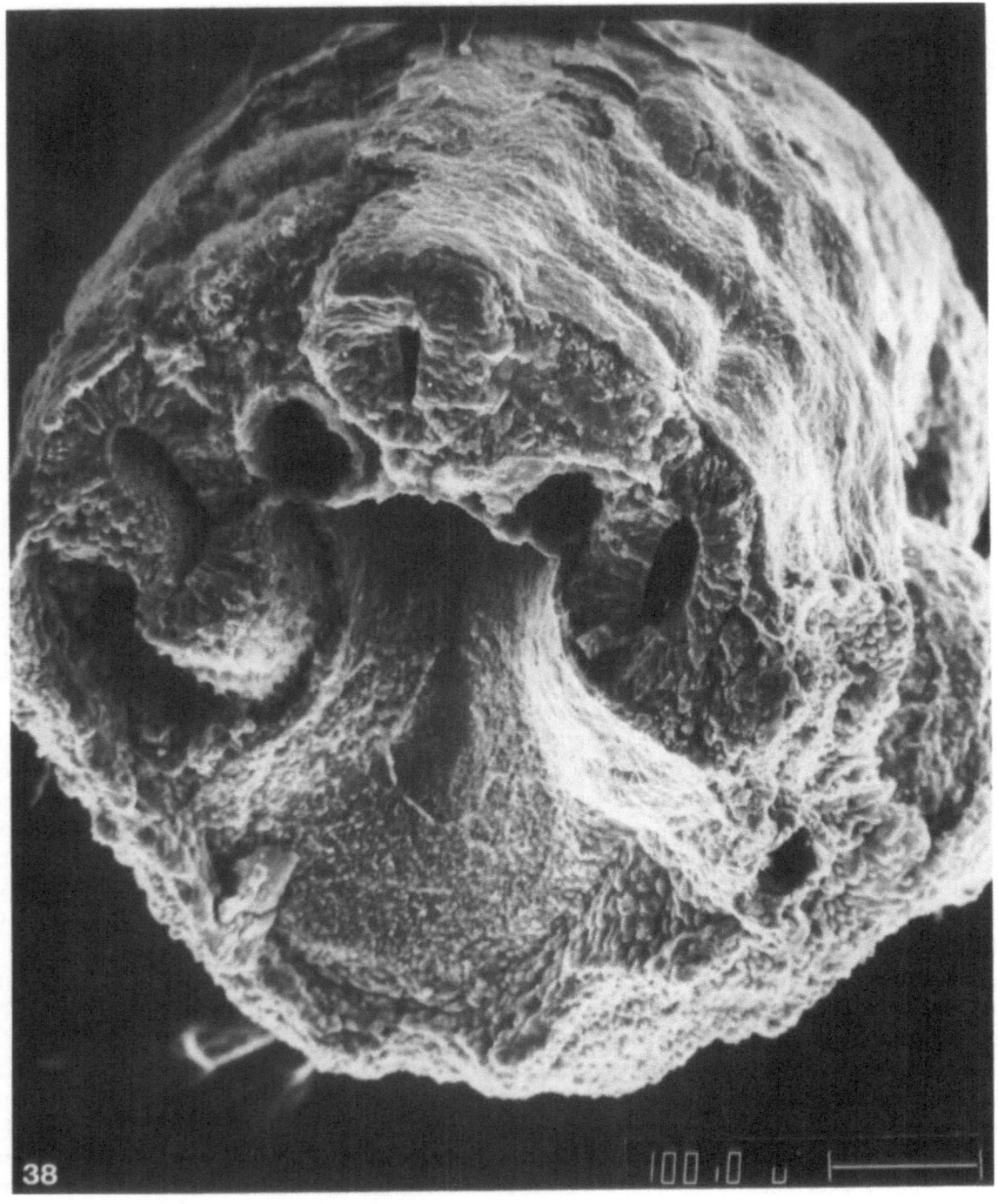

Figure 38. Transverse dissection of a 13-somite embryo at the level of somite 4. Stage 6-2. Structures depicted: surface ectoderm, medullary tube, notochord, somites, paired dorsal aortae, lateral mesoderm: somato- and splanchnopleurae, anterior gut, and yolk sac (anterior portion).

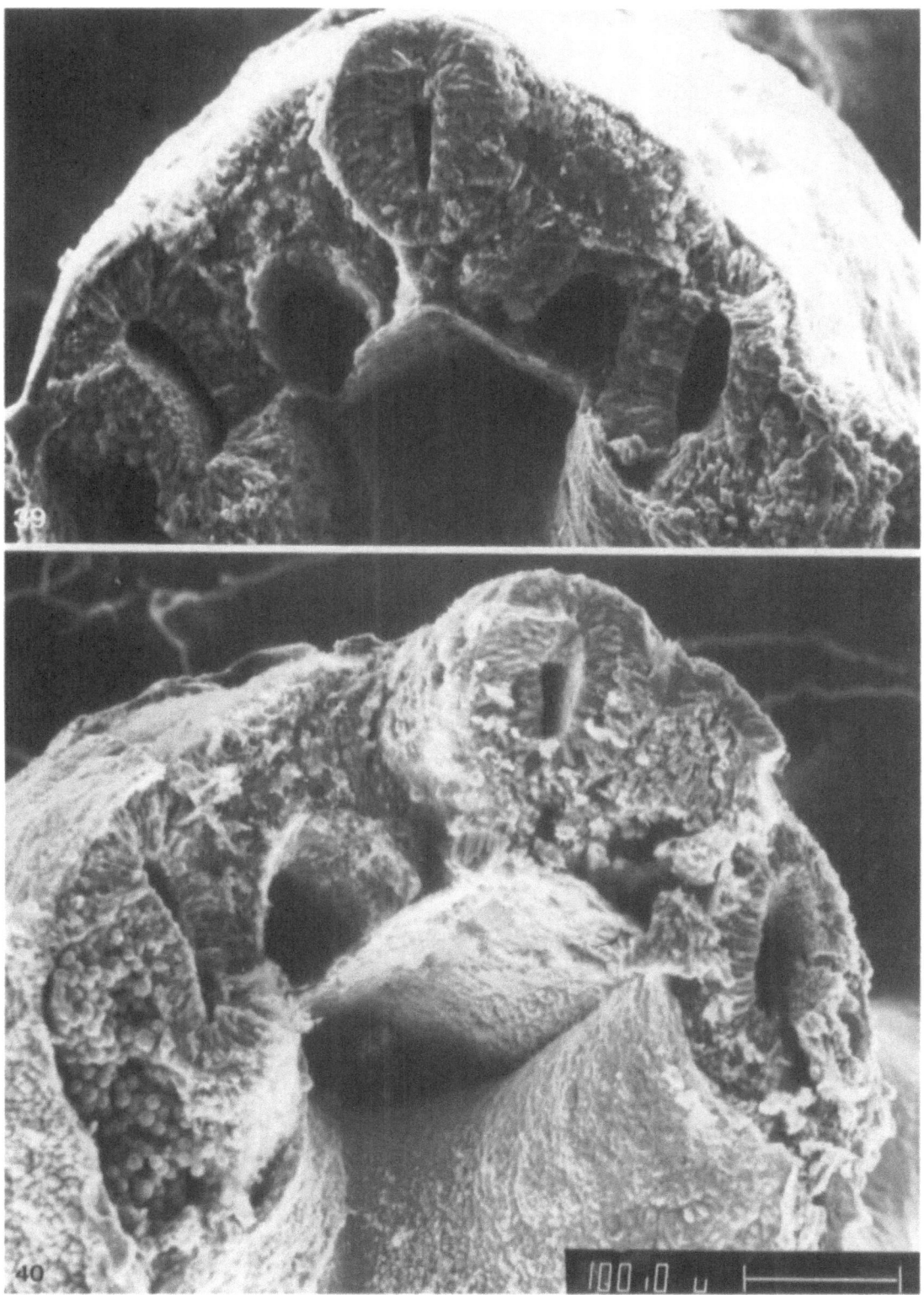

Figures 39 and 40. Germ layers in a dissected 13-somite embryo. Stage 6-2. Structures depicted: medullary tube, notochord, two dorsal aortae, lateral mesoderm: somato- and splanchnopleurae, anterior gut. Vitelline vein with erythroblasts adjacent to lateral mesoderm (somatopleura) on the left side (Fig. 40).

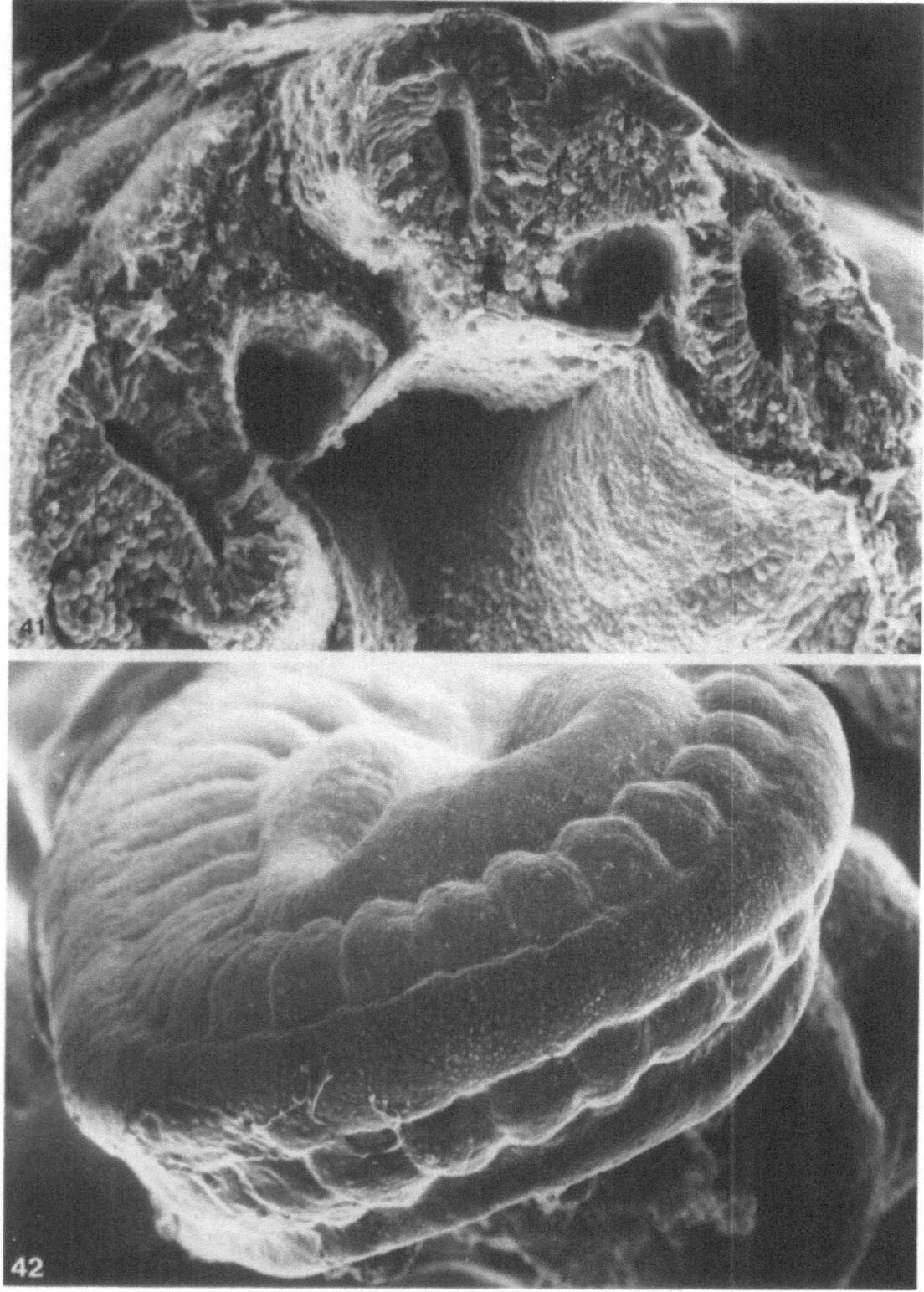

Figures 41 and 42. (41) Germ layers in a 13-somite embryo. Stage 6-2. Note the proliferation of mesodermal cells from the sclerotome of the somite located on the right side. **(42)** Somites – mesodermal vesicles – located along a centrally situated medullary tube. The mesodermal structures are covered by surface ectoderm.

detached from the 'primitive ectoblast' at the bi-laminar stage colonizing the endoblast. The mammalian mesoderm is detached from the primitive ectoderm (ectoblast) by way of a primitive groove. The notochord is invaginated from the primitive node. The invagination of the notochord is probably the most conservative trait in vertebrate blastogenesis. The mesoderm (mesoblast) is the main source of mesenchyme. In addition to the mesoderm, the mesenchyme is contributed to by neuro-ectodermal cells from the closing neural tube and the neural crest (ectomesenchyme). The mesenchyme is able to differentiate into the connective tissue, angioblasts, hermatopoetic tissue, and smooth muscle cells, and is involved in the morphogenesis of all organs. The following list enumerates the main organs according to the involvement of ectoderm, endoderm, mesoderm, and mesenchyme in their morphogenesis.

1) Organs with an ectodermal component
   Surface ectoderm
   a) skin (epidermis), hair, nails, sweat glands, sebaceous glands, mammary glands
   b) ectoderm of primitive oral cavity: hypo-physeal pouch–adenohypophysis, enamel of teeth, lining of nasal cavity
   c) lenses of eyes, corneal anterior epithelium, conjuctival epithelium
   Neural ectoderm
   a) olfactory and auditory sensory epithelium
   b) central nervous system and retina
   c) peripheral nervous system and adrenal medulla

d) ectomesenchyme: meninges, melanocytes, odontoblasts, some structures related to pharyngeal arches

2) Organs with an endodermal component
   a) pharynx, esophagus, stomach, intestines, thyroid, parathyroid, thymic reticulum, pancreas, liver and other glands of digestive tract
   b) middle-ear cavities and pharyngotympanic tubes
   c) respiratory organs: larynx, trachea, bronchi, lung alveoli
   d) urogenital sinus derivatives: urinary bladder (except the trigone), female urethra, ca. 1/5th of vagina, intramural and upper prostatic portions of male urethra, prostatic glands

3) Mesodermal and mesenchymal derivatives
   a) of somitic origin: skeletal musculature
   b) from somatopleura: connective tissues, cartilages, bones, synovial membranes
   c) from splanchnopleura: connective tissue, cartilages, bones, synovial membranes, visceral mesenchyme, visceral musculature, hematopoetic tissue, heart, spleen
   d) from neural crest: meninges, mesenchyme of pharyngeal arches, odontoblasts
   e) from somato- and splanchnopleurae: pleural, pericardial and peritoneal linings, endothelial cells, vessels
   f) from intermediate mesoderm (uromesoderm): kidneys
   g) special mesoderm–mesenchymal organs: cortex of adrenals, gonads (except germ cells), derivatives of paramesonephric ducts: uterine tubes, uterus, ca. 4/5 of the vagina.

4. MESENCHYMAL AND MESODERMAL ORGANS

The development of the mesenchyme (Figs. 11c, 12c, 13c, 15c and 42–48)

Connective tissues, the vessels, the heart, the blood, and organs of locomotion (such as the skeleton, the bands, and the cross-striated muscles) originate from mesoderm. The mesoderm consists of segmentally arranged somites (dorsal portion of the mesoderm), somitic stalks (intermediary portion of the mesoderm), and nonsegmented somatopleura and splanchnopleura (lateroventral portion of the mesoderm). The cells of the mesoderm in early somite embryos temporarily exhibit an epitheliallike organization. In contrast to the true epitheliums, the basal portions of mesodermal cells extend into processus attaching to neighboring ectodermal and endodermal structures. The cavity formed within the mesoderm, between the somatopleura and splanchnopleura, is the coelom. The cavities present within the somites constitute the myocoele. Stellate cells migrating into spaces between germ layers constitute the mesenchyme, most of which is of mesodermal origin. Some mesenchymal cells, especially around the CNS and in the branchial arches, are of neuroectodermal origin.

In human embryos, 42–44 somites are formed. With the exception of 3–5 cranial segments that transform directly into mesenchyme, three portions are distinguishable on each somite: the dorsolateral portion represents the myotome, the ventromedial portion is the sclerotome, and the ventrolateral portion is the dermatome. Myotomes represent primordia of skeletal muscles. Sclerotomes contribute to mesenchymal cells migrating to the notochord, transforming into mesenchymal blastemas of the vertebrae. The cells of the lateral portion of somites (dermatomes) contribute to the mesenchyme located under the surface ectoderm. Cells migrating from the somatopleura contribute to the mesenchyme of the lateral and ventral portions of the body wall. Cells migrating from the splanchnopleura give rise to the mesenchyme accompanying the endodermal digestive tube and its derivatives. Condensations of mesenchymal cells are known as blastemas: the chondrogenic blastema precedes differentiation of the cartilages, the desmogenic blastemas precede development of the bands and the tendons, the myogenic blastemas give rise to the musculature.

The notochord (Figs. 8c, 13c, 15c and 49–52)

The chordomesodermal canal, notochordal plate, and notochord

The chordomesodermal canal is incorporated primarily within the ceiling of the yolk sac and extends from the primitive pit of the primitive node to the prechordal plate. As the ventral wall of the chordomesodermal canal disintegrates, the canal changes into the notochordal plate. The notochordal plate and the neural plate are temporarily in direct contact, which determines the development of the ventral plate of the CNS.

As the amount of the intraembryonic mesenchyme increases, the notochordal plate becomes detached from the CNS as well as from the endoderm of the gut and transforms into a notochord with a distinct capsule. The notochord is located between the CNS and the gut. Mesenchyme accumulating around the notochord becomes condenses into the mesenchymal primordia of vertebrae and into the intervertebral disks. The rostral end of the notochord ends at the posterior wall of the Rathke's pouch.

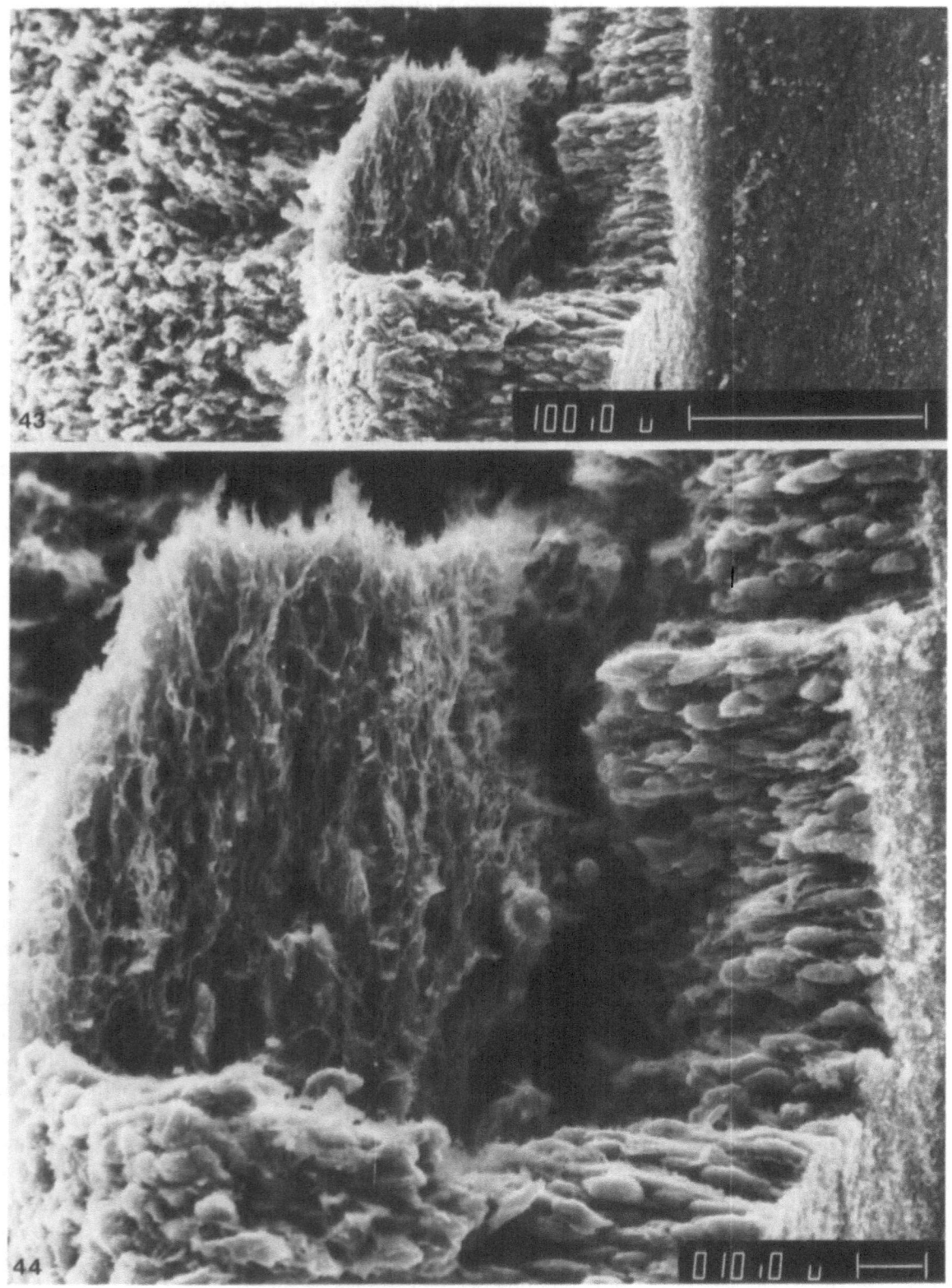

Figures 43 and 44. (43) Dissected somite with a laterally located myotome (right) and sclerotome left. **(44)** Densely packed collagen fibers formed by mesodermal cells precede the differentiation of myoblasts.

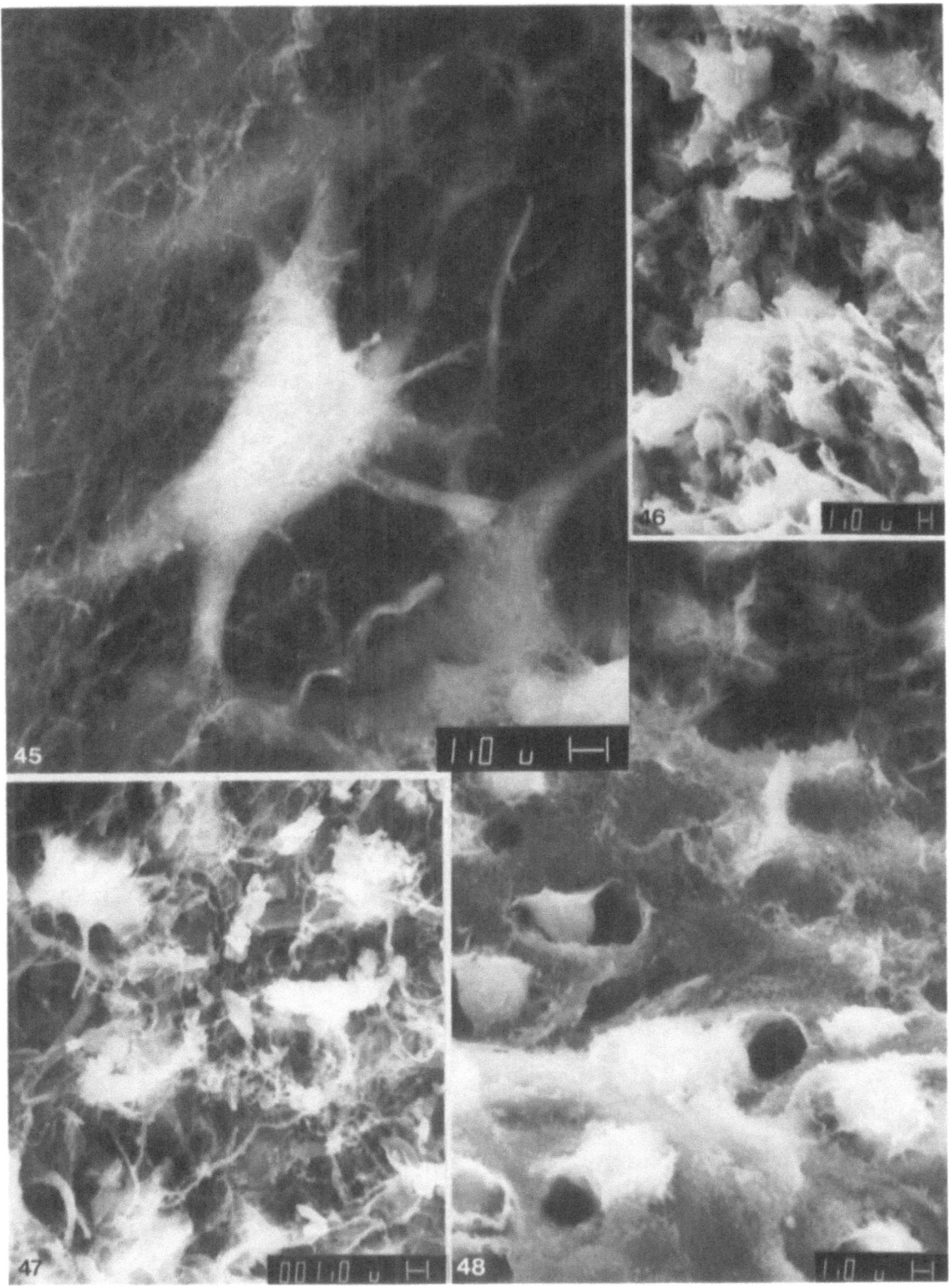

Figures 45–48 (45) Stellate mesenchymal cells and extracellular fibers. **(46)** Cells of a mesenchymal blastema preceding formation of cartilage. **(47)** Mesenchymal connective tissue. **(48)** Transformation of mesenchyme into a hyaline cartilage. Note the organization of the intercellular substance and formation of capsules around the chondrocytes.

42

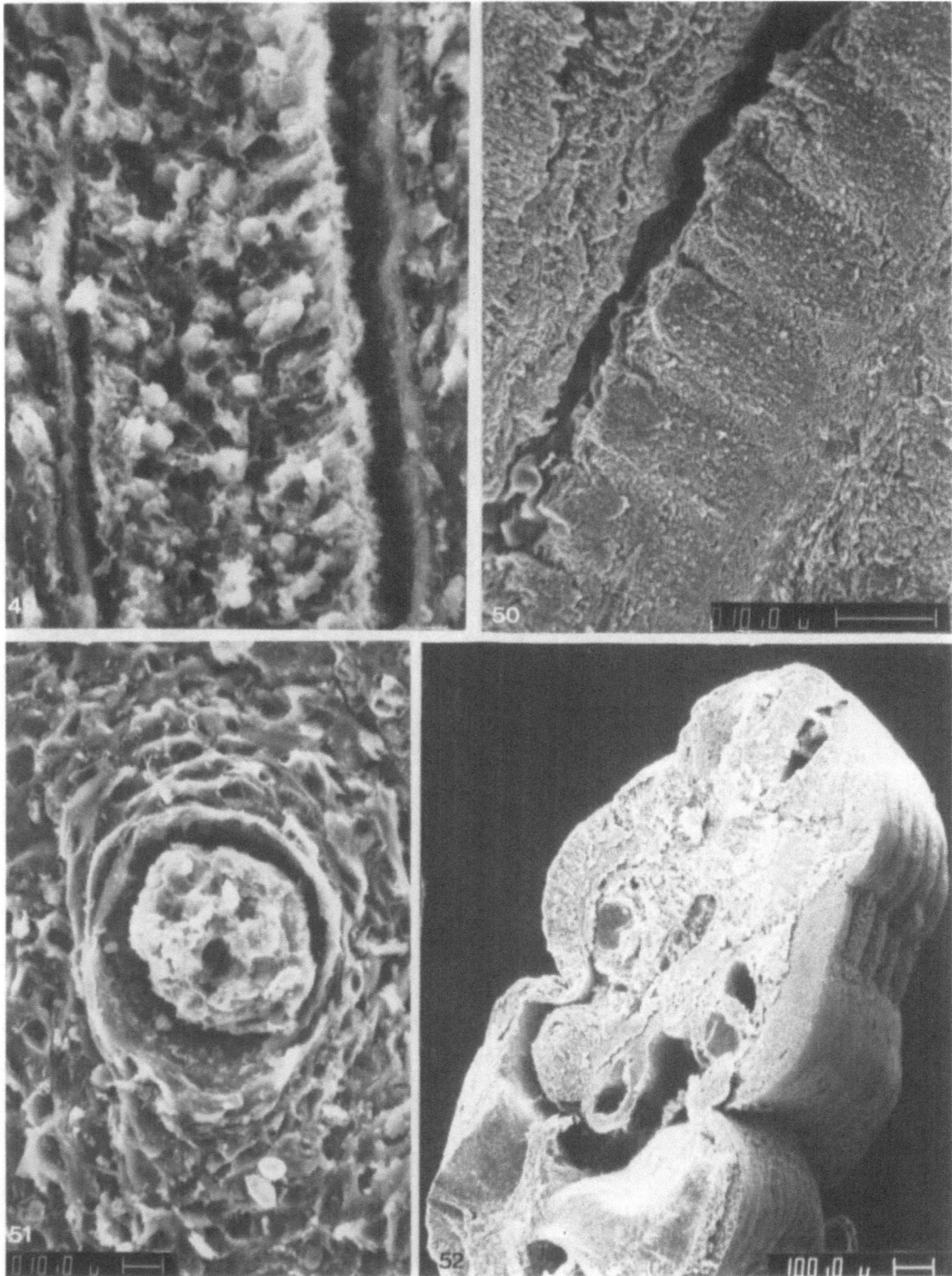

Figures 49–52. (49) Longitudinal dissection of the notochord. The capsule is artificially detached from the cells. **(50)** Formation of the vertebral column. Precartilaginous vertebral bodies are separated by dense intersegmental mesenchyme. **(51)** Remnant of notochord within the cartilaginous primordium of the vertebral body. **(52)** Transverse dissection of an embryonic body. Stage 7-4. Mesenchyme is the 'main substance' of the embryo. Epithelial structures are incorporated within, or supported by, mesenchyme. The surface is covered by ectoderm.

The skeleton

Two types of bones may be distinguished according to their types of ossification. In chondrogenic bones, the development of the bone is preceded by a cartilaginous model. The cartilage is resorbed, and replaced by bony tissue through endochondral and perichondral ossifications. The desmogenic bones are formed by desmogenic (membranous) ossification. In desmogenic ossification, the mesenchymal cells arrange in sheets or rows in the vicinity of inducing structures (such as on the surface of the primitive dura mater of the brain) and produce the prebone (osteoid) material that undergoes mineralization.

The axial skeleton (Figs. 41C and 53)
The axial skeleton is formed in a close relationship to the notochord. This portion of the notochord participating in the morphogenesis of the base of the skull is incorporated within the endoderm of the pharynx, and the chondrogenic mesenchyme of the skull base condenses between the notochord and the brain. This portion of the notochord, which is located between the medullary and digestive tubes, represents an axial structure around which the vertebral bodies are formed. During formation of the primordium of the vertebral body, each mesenchymal segment corresponding to one sclerotome divides into a cranial portion and a caudal portion separated by an intervertebral fissure. Consequently the caudal portion of sclerotome joins the cranial portion of the following sclerotome. While the myotomes remain limited to the original segments, the vertebral bodies are intersegmental. The muscles connect neighboring vertebrae, which enables movement in the intervertebral joints. The cartilaginous blastema preceding the development of each vertebral body develops mainly from the caudal portion of the intervertebral disk. Paired neural and costal processus grow from the vertebral primordium. Neural processus join dorsally to the medullary tube and fuse into the vertebral arcus. Each arcus has a dorsally oriented spinal (laminar) process. In addition, a transverse process and a cranial and caudal articular processus appear on each neural process. Costal processus of vertebral primordia grow laterally (in lower vertebrates, within the intersegmental septa), and in the thoracic area represent primordia of the ribs. The free ends of ten cranial ribs join together and the cartilages formed between cranial ribs 1–7 fuse in the midline anterior to the heart into a sternal primordium.

Primordia of lumbar vertebrae are characterizd by fusion of the lateral processus (formed on the neural processus) with the costal processus into costal transverse processus. In the five sacral vertebral primordia, costal processus fuse with the vertebral bodies and the transverse processus fuse craniocaudally into the lateral portions of the sacral bone. The neural (laminar) processus of the caudal sacral vertebrae fail to form a complete neural arch. In this way, the hiatus of the sacral canal is formed.

The blastematous primordia of the five coccygeal vertebrae fuse into the coccygeal bone. Their neural processus, which fail to join dorsally, represent horns of the coccygeal bone.

The skull (Fig. 42C)
The skull consists of two portions: the neurocranium, formed around the brain; and the viscerocranium, located around the most cranial portion of the digestive tube. The neurocranium consists of a chondrogenic portion belonging to the cranial portion of the axial skeleton and of cartilages related to the otic and nasal placodes. The desmogenic portion of neurocranium is contributed to by the desmogenic bones of the calvarium. The viscerocranium (skeleton of the face) comprises desmogenic bones ossifying around the cartilaginous skeleton of the two cranial branchial arches.

In relation to the notochord, the mesenchyme contributing to the chondrocranium is classified as parachordal (in fact, retrochordal) and prechordal. The parachordal mesenchyme is located between the dorsal pharyngeal wall (containing the notochord as a midline structure) and the base of brain vesicles up to the Rathke's (hypophyseal) pouch. The prechordal mesenchyme occupies an area located anterior to the Rathke's (hypophyseal) pouch, reaching the olfactory placodes. The parachordal mesenchyme (or basal plate) connects with the cartilaginous otic capsules. The prechordal portion of mesenchyme fuses with the cartilaginous nasal capsule. The parachordal plate with sclerotomes of 4–5 disappearing occipital somites contributes to the occipital bone. The neural processus of the segments incorporated in the occipital bone extend

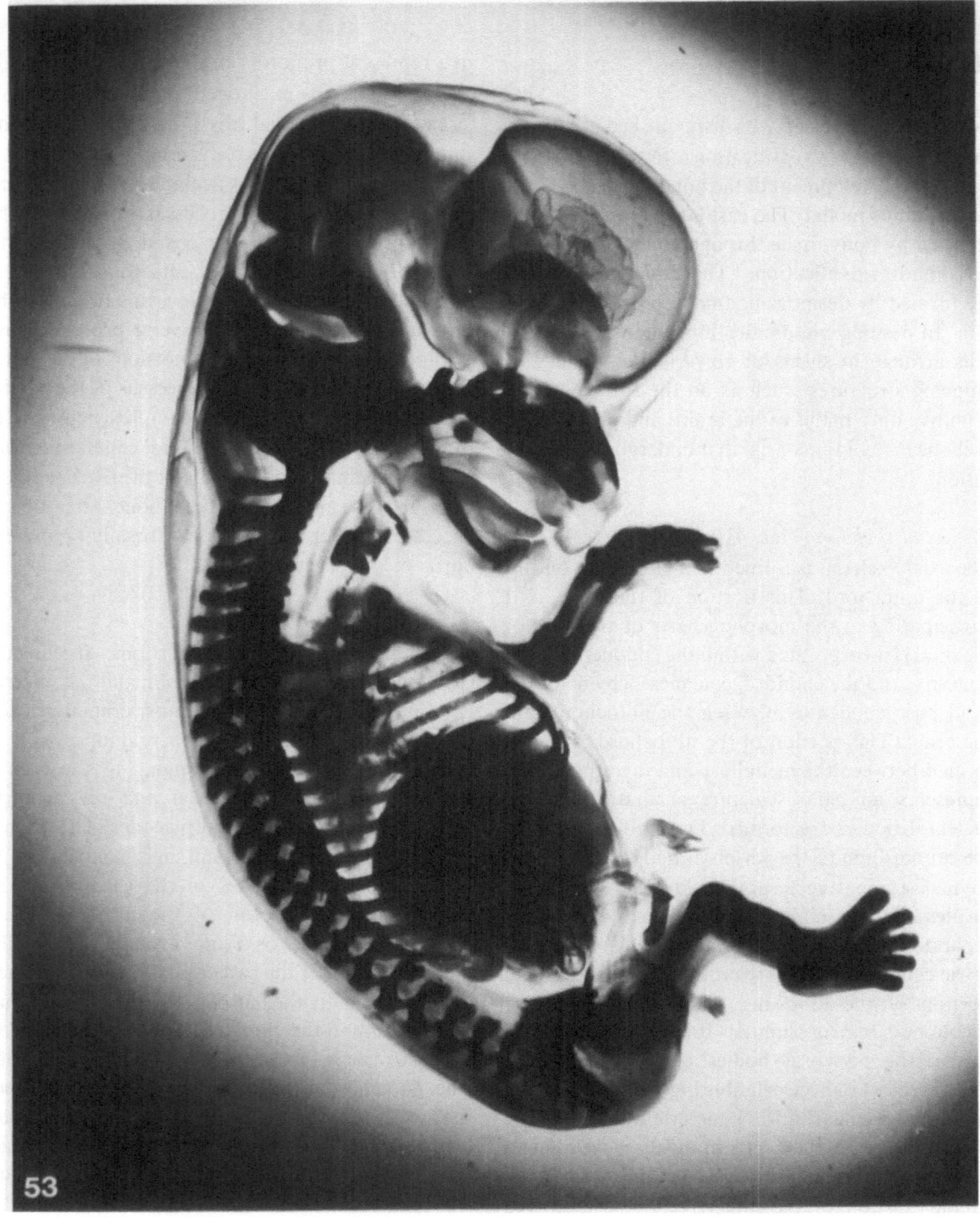

Figure 53. Cleared embryo, right half. Stage 8-2. Cartilaginous skeleton visualized by alcian blue stain.

dorsally around the medullary tube, leaving the foramen magnum and forming the lower portion (tectum posterius) of the occipital squame. The upper portion of the occipital squame is desmogenic in origin. The otic capsules are composed of a dorsolateral, canalicular portion and a ventromedial, cochlear portion. The otic capsule contributes to the petrous and mastoid parts of the temporal bone.. The corpus of the sphenoid bone develops from mesenchyme located around the Rathke's pouch, which fuses with orbitosphenoid and alisphenoid cartilages contributing to the ala orbitalis (ala parva) and ala temporalis (ala magna) and lateral pterygoid processus of the sphenoid bone.

The prechordal mesenchyme differentiates into a vertical nasodorsal plate (interorbital septum and nasal septum) and into two nasozygomatic plates or paranasal cartilages. The paranasal cartilages and the vertical plate fuse into the nasal capsule. The ethmoid bone and the nasal conchae ossify within the nasal capsule.

The following bones of the neurocranium are of desmogenic origin: the upper portion of the occipital squame, the squame of the temporal bone, the parietal bone, the frontal bones, and the lamina medialis of the pterygoid process of the sphenoid bone. The viscerocranium consists of desmogenic bones. The premaxillae, the vomer, the lacrimal bones, the nasal bones, and the zygomatic bones originate from the nasofacial mesenchyme. The maxillary and palatinal bones develop from the maxillary portions of the first pharyngeal arch. The mandibula ossifies lateral to the Meckel's cartilage of the first pharyngeal arch. In addition, mesenchyme of the dorsal portion of the first pharyngeal arch gives rise to the cartilaginous primordia of the malleus and incus. Mesenchyme of the second pharyngeal arch (the Reichert's cartilage) transforms dorsally into the cartilaginous primordium of the stapedial crura and contributes to the styloid process of the temporal bone. The middle portion of the arch gives rise to the stylohyoid ligament. The ventral portion of the arch changes into the lesser horns, and into the upper portion of the body of the hyoid bone. The greater horns and the lower portion of the hyoid bone are formed from the cartilaginous portion of the third pharyngeal arch.

The appendicular skeleton (Figs. 43c–45c)

The primitive limb buds contain unsegmented lateral mesenchyme growing from the somatopleura. This mesenchyme is invaded by myoblasts and nerves from the adjacent segments.

Within the primordia of the extremities, the mesenchyme giving rise to the skeleton condenses into the chondrogenic blastematous primordia of bones (the blastematous stage). From the chondrogenic blastemas, hyaline cartilage models of bones (the cartilaginous stage) are formed that are transformed into bones by ossification. Ossification and the related growth of bones proceeds for years by endochondral and perichondral ossifications. The endochondral ossification results in spongy bone and perichondral ossification results in compact bone.

The primary ossification center is located in the cartilaginous diaphysis. The primary ossification proceeds gradually to the ends of diaphysis. The epiphysis remains cartilaginous for a considerable time. Secondary ossification centers appear within the epiphysis. Spongy bone is formed from center to periphery. A cartilaginous epiphyseal plate remains temporarily between the diaphyseal parts of the bone. The epiphyseal plate makes possible the longitudinal growth of the bone.

The joints and bands

Joints are preformed between the blastematous primordia of bones. As the cartilage of the model bone becomes evident, the mesenchyme within the future joint cavity becomes preserved. Condensed mesenchymal perichondrium becomes evident around the cartilaginous models of bones. The perichondria of the articulating cartilaginous models of bones connect, overbridging the future joint cavity and contributing to a connective tissue capsule around the future joint cavity. The loose mesenchyme within the joint cavity disappears during the period of ossification of bones participating in the articulation. The ligaments are formed from mesenchymal connective tissue.

The cross-striated musculature

The striated (skeletal) muscles develop from myotomes or from closely related mesenchymal material. Myotomes are composed of elongated cells that synthesize cross-striated muscle fibers. The myotomes are innervated by nerve fibers of the adjacent segments of the medullary tube. Cells growing from the myotomes extend ventrally and penetrate between the surface ectoderm and the somatopleura. Consequently the ventral extensions of myotomes known as hypomeres become separated from the dorsal portions of myotomes, or epimeres, by an intermuscular septum (myoseptum). The myoseptum separates the dorsally located epaxonic from the ventrally located hypaxonic musculature. The epaxonic portions of myotomes are innervated by the dorsal primary ramus of the spinal nerve, the hypaxonic portions by the ventral primary ramus of the nerve. The muscles originating from epimeres contribute to the extensor muscles of the spine separated from the musculature originating from hypomeres by a fibrous sheath known as the deep layer of the lumbosacral fascia. The lumbosacral fascia is related to the intermuscular septa of cervical, thoracic, and lumbar myomeres.

During morphogenesis of the skeletal muscles, each myotome forms several portions and layers. Portions of different myotomes may split, migrate, and become reoriented. The myotomes of those segments that were incorporated during phylogenesis into the skull are related to the extrinsic ocular muscles, to the pharyngeal arches, and to the tongue. The hypothetical three preotic myotomes anterior to the ear placodes provide musculature innervated by the IIIrd, IVth, and VIth nerves. The musculature related to the first pharyngeal arch (the masticatory muscles) is innervated by the Vth nerve. The musculature related to the second pharyngeal arch (the mimic muscles of the face) is innervated by the VIIth nerve. The lingual musculature is related to the four occipital segments. The musculature of the proximal limb is derived from four caudal cervical and two cranial thoracic segments. Musculature of the distal limb relates to the five lumbar and two cranial coccygeal segments. Although developmental changes in the morphogenesis of different skeletal muscles may be complicated, the segmental origin of musculature can be always traced by the segmental innervation (Figs. 54–64).

The cardiovascular organs and hematopoesis

The cardiovascular organs – the heart, the vessels, and the hematopoetic tissue – are mesenchymal derivatives. First primordia of vessels appear as angiogenic clusters in the mesoderm of the yolk sac and as angiogenic cords in the mesenchyme of the early chorion. The angiogenic clusters of the yolk sac represent the first hematopoetic tissue, known as the blood islands. Peripheral cells of the blood islands flatten and transform into the endothelial cells that interconnect and form the network of yolk-sac capillaries. The centrally located cells of the blood islands, the hematopoetic stem cells, detach and develop into the first blood cells of the embryo (Figs. 65–68).

Within the body of the embryo, the first angiogenic cords appear in the splanchnic mesoderm between the pericardial cavity and the endoderm of the anterior gut, contributing to the cardiogenic plate. Other angiogenic cords are formed within mesenchyme underneath the somites and represent primordia of the dorsal aortae, or connect the capillaries of the yolk sac and the heart plate as primordia of vitelline veins; angiogenic cords in the connecting stalk form primordia of the umbilical vessels.

The heart (Figs. 69–93)
Three consecutive stages are distinguished in the morphogenesis of the heart:

1) Constitution of the pericardial cavity and the heart tube.
2) Formation and differentiation of the heart loop.
3) Septation of the heart loop.

The pericardial cavity and the heart tube. The cardiogenic plate is a horseshoe-shaped mesoblastic condensation anterior to the rostral end of neural groove. The plate transforms into a primitive horseshoe-shaped pericardial cavity lined by a mesodermal splanchnopleural layer. Myoepicardial cells detached from the left-sided and right-sided

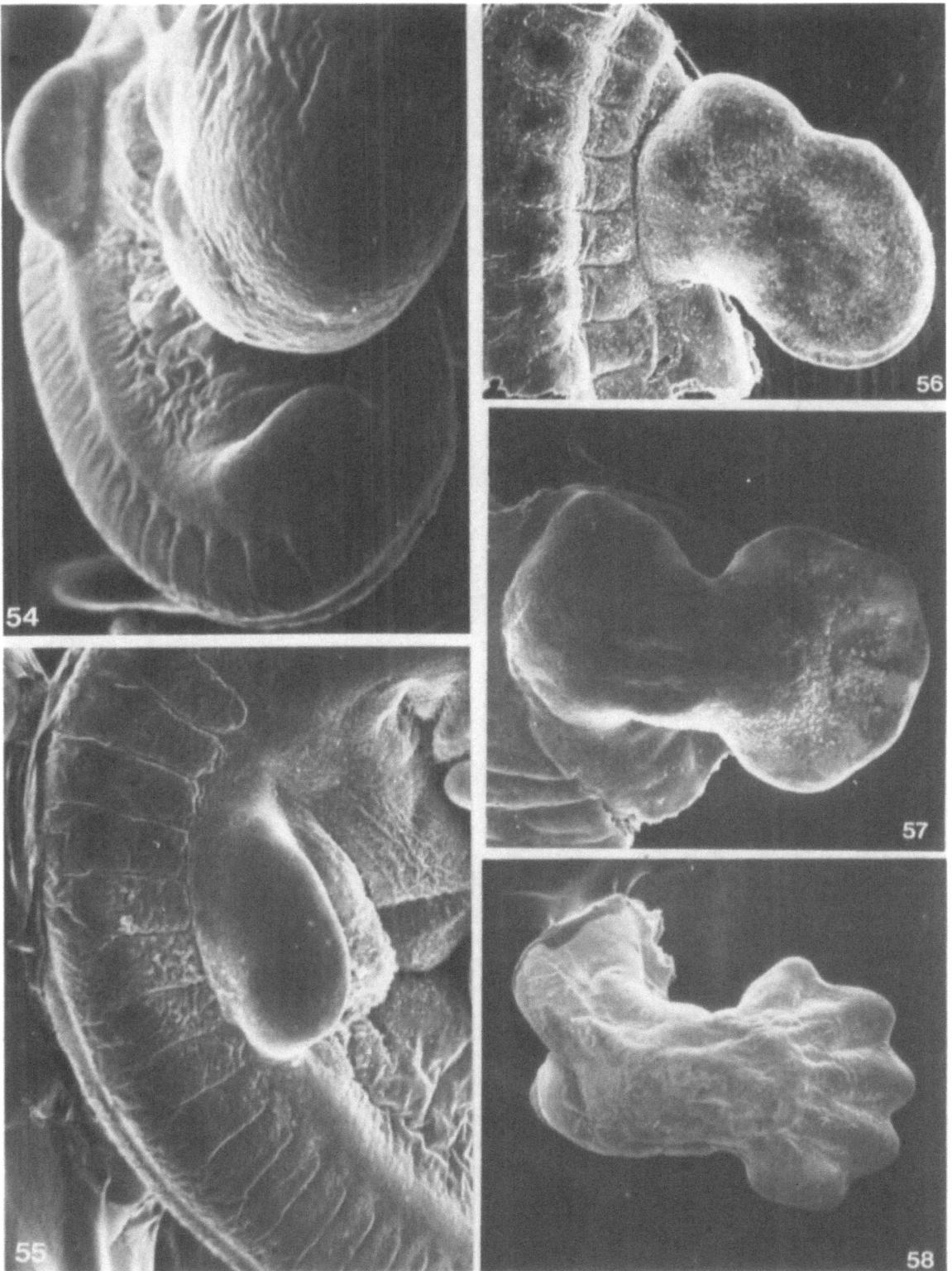

Figures 54–58. Formation of the limbs. **(54)** Limb buds. Stage 7-2. Forelimb buds are more advanced than the hindlimb buds. **(55)** Unsegmented forelimb bud located in the lateral mesenchyme ventrally to four cervical and two thoracic segments. Stage 7-2. **(56)** Bisegmented forelimb bud with an ectodermal apical ridge. Stage 7-4. Segmentation of the dorsal mesoderm is still preserved. **(57)** Bisegmented forelimb with finger rays on the hand plate. Stage 7-5. Segmentation of the dorsal mesoderm has disappeared. **(58)** Trisegmented forelimb with a distinct stylopodium, zeugopodium, and autopodium with finger tubercles. Stage 7-6.

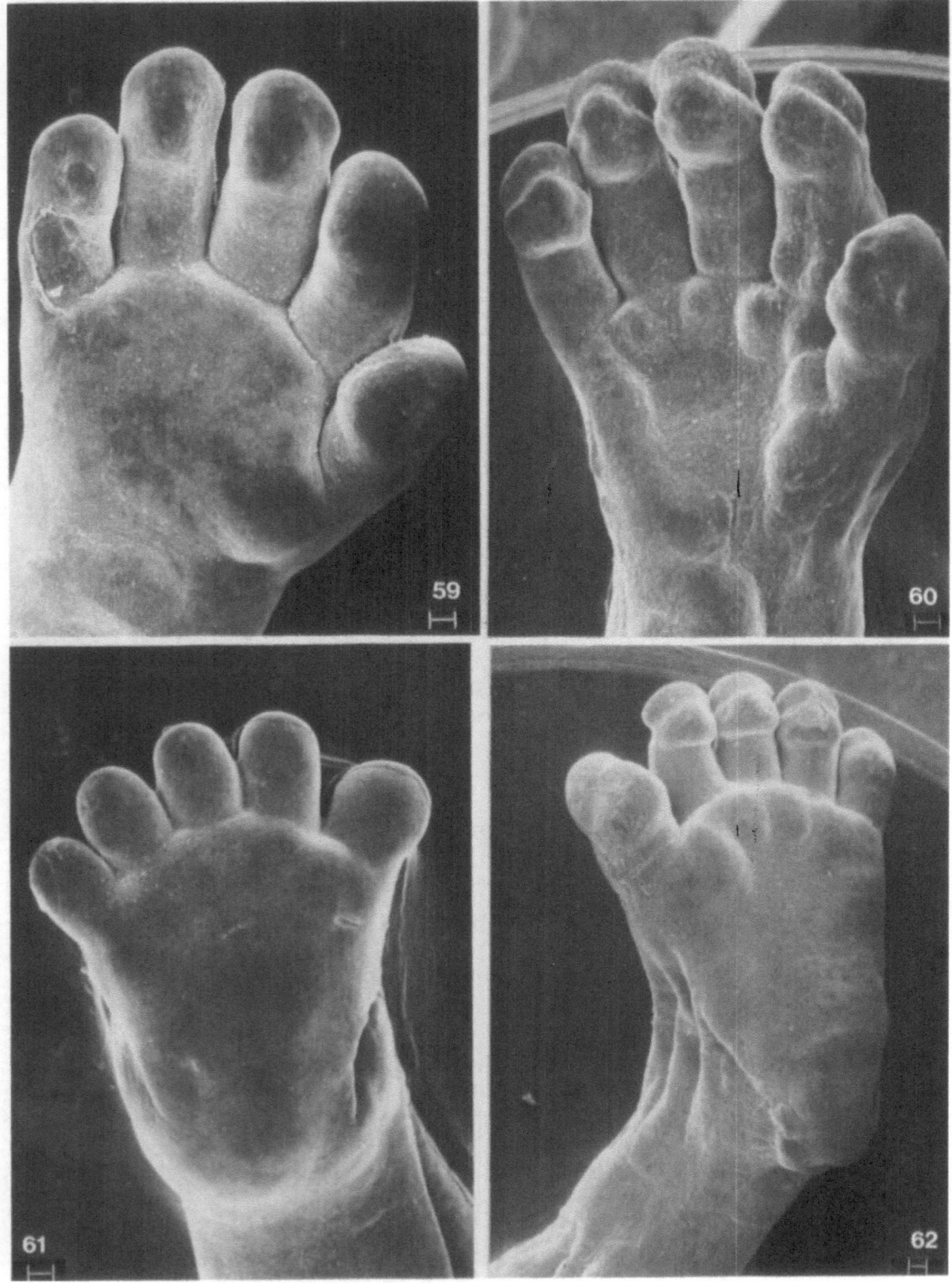

Figures 59–62. **(59)** Differentiated hand with fingers. Stage 8-1. **(60)** Differentiated hand; fingers with volar pads. Stage 8-2. **(61)** Foot with differentiated toes. Stage 8-1. **(62)** Foot; toes with volar pads. Stage 8-2.

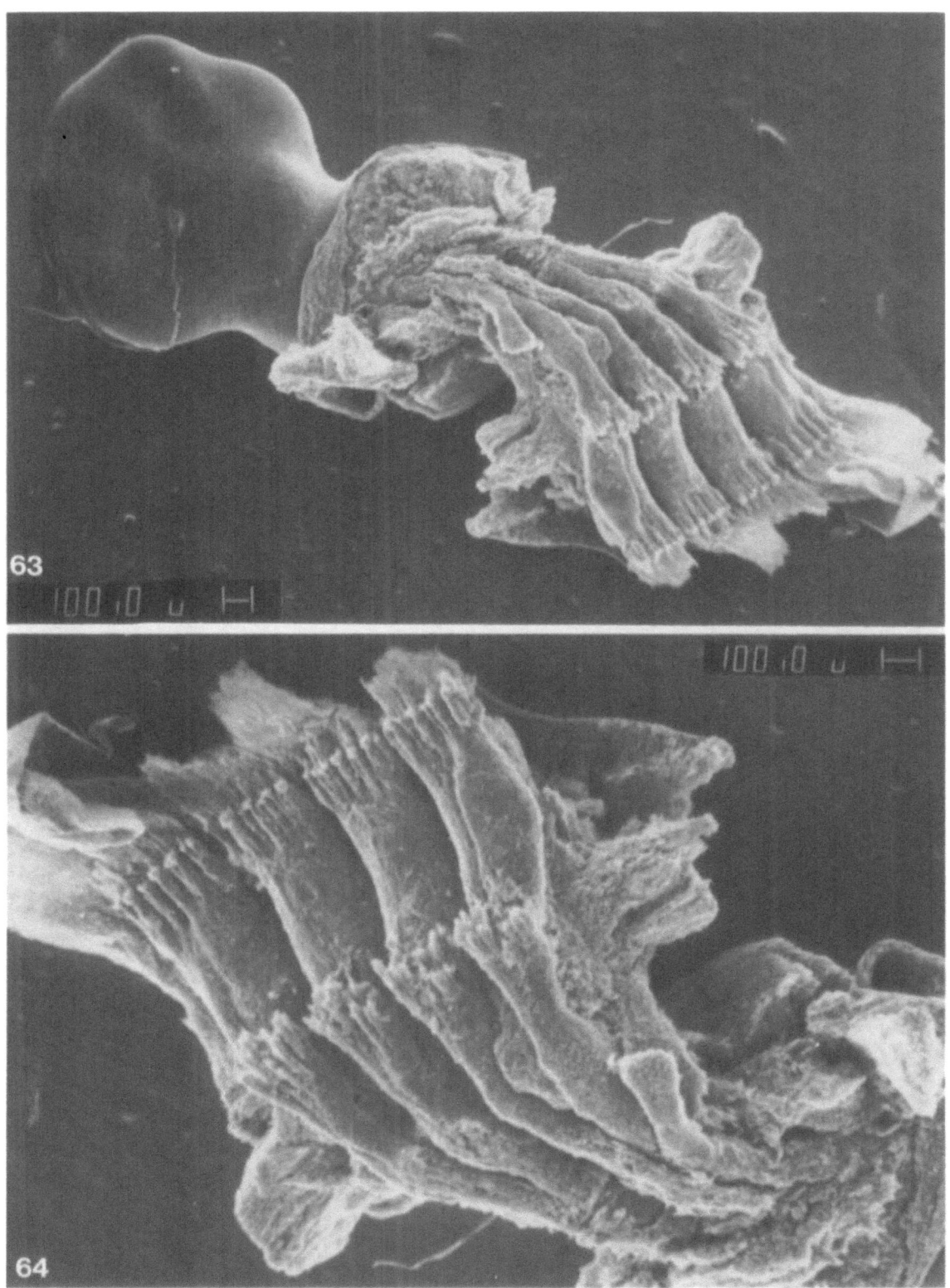

Figures 63 and 64. Innervation: spinal ganglia and nerve stems of the upper right limb. Stage 7–5. **(63)** Ventral view. **(64)** Medio-inferior view.

50

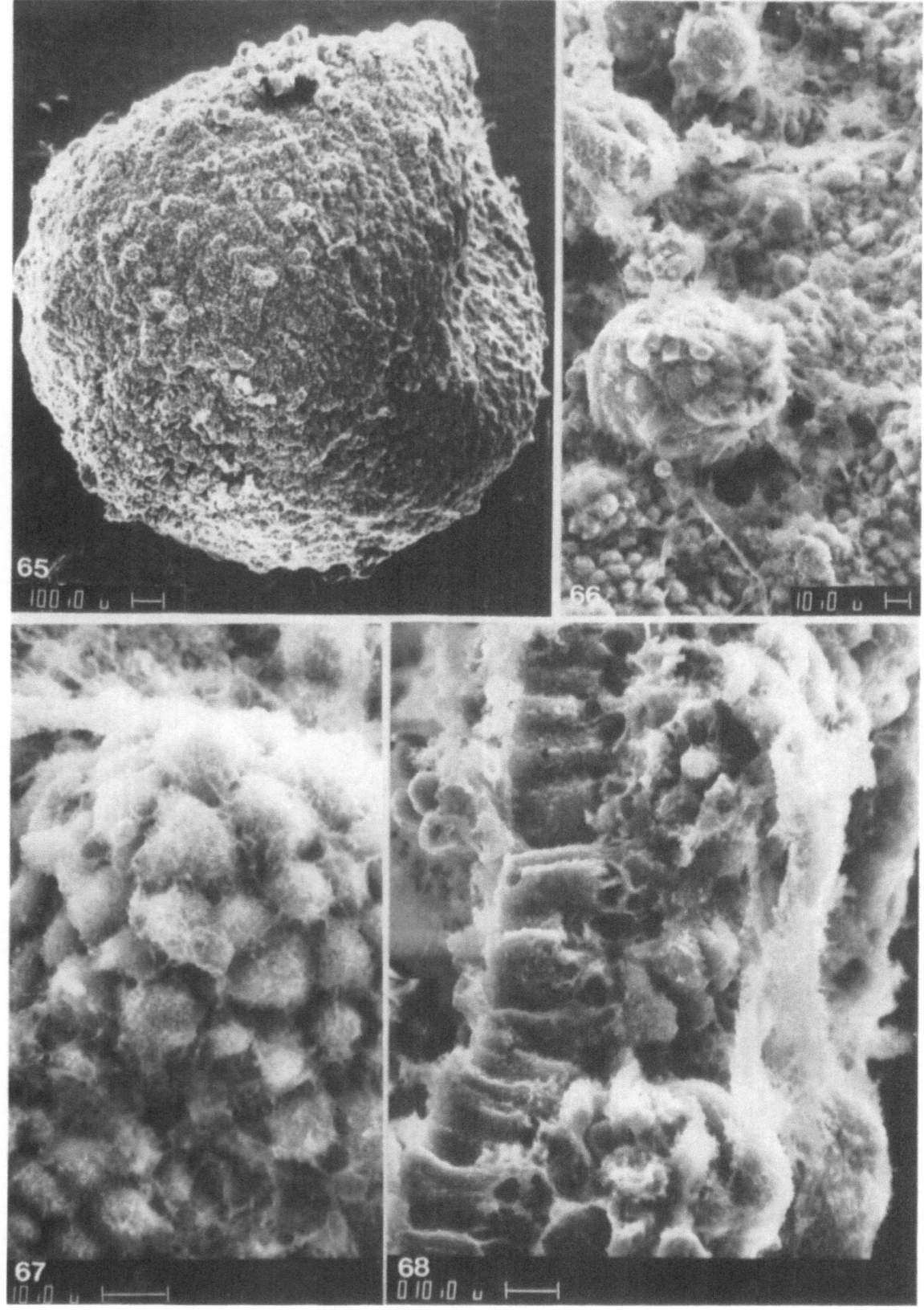

Figures 65–68. (65) Yolk sac. Stage 6-3. SEM **(66)** Surface of the yolk sac. Stage 6-3. Mesoderm with blood islands. SEM **(67)** Surface cells of a blood island. **(68)** Dissected wall of the yolk sac. Stage 6-3. Endodermal cylindrical cells with cytoplasmic granules, primitive blood cells, and cells of yolk sac mesoderm. SEM.

splanchnopleural layer constitute paired tubes that consequently fuse, forming the heart tube. The pericardial cavity, containing the heart tube, descends underneath the foregut. The heart tube protrudes into the pericardial cavity suspended by the dorsal mesocardium. A ventral mesocardium is never formed in man. The wall of the heart tube consists of an endothelial tube and a myoepicardial mantle separated by the cardiac jelly. Beginning from the caudal, venous end of the primitive heart tube, the following segments can be recognized: a paired venous sinus, an atrial segment, and a bulboventricular segment.

Segmentation of the heart loop (Figs. 69–79). The heart tube located within the pericardial cavity changes in relation to the longitudinal growth into an S-shaped loop on which the following three segments become evident: a sinoatrial portion (located dorsally and cranially on the right side, comprising the venous sinus and atrial segments of the primitive heart tube) a ventricular portion (the descending portion of the loop located on the left side), and a bulbar portion (ascending portion of the loop). The sinoatrial portion of the heart loop represents the primordia of both atria. The ventricular portion gives rise to the trabecular portion of the left ventricle. The bulbar portion becomes divided into (a) the ventricular segment – the primordium of the trabecular portion of the right ventricle, (b) the conus – the primordium of smooth portion of both ventricles, and (c) the aortico-pulmonary segment – the primordium of the proximal portion of both – the aorta and the pulmonary artery. The sinoatrial portion and the ventricular portion of the loop are separated by the atrioventricular constriction. The narrow portion of the heart tube in the area of the atrioventricular constriction is known as the atrioventricular canal. The bulboventricular constriction is evident from outside as the bulboventricular sulcus and from inside the heart tube as the bulboventricular fold.

Septation of the heart (Figs. 80–83). The septation of the embryonic heart involves:

a) Incorporation of the venous sinus into the right atrium and separation of the right and left atria by the septum primum and the septum secundum.

b) Division of the atrioventricular canal into right and left atrioventricular openings and the formation of the tricuspid and bicuspid valves.

c) Formation of the interventricular septum.

d) Development of the aorticopulmonary septum.

e) Formation of the semilunar valves.

a) The primitive venous sinus consists of a transverse portion and the right and left sinus horns. The venous sinus is separated from the atrium by the sinoatrial fold. Each horn recives the omphalomesenteric (vitelline) vein, the umbilical vein, and the common cardial vein. The left-sided veins obliterate and the left-sided horn of the sinus transforms into the transverse portion of the coronary sinus and into the oblique vein of the heart (vein of Marshall).

The right-sided horn and the rest of the sinus are incorporated into the expanding right atrium and transform into the smooth portion of the atrial wall. The trabeculated portion originates from the atrial segment of the heart loop. The sinoatrial orifice is demarcated by the right and left venous valves fusing cranially into a septum spurium. As the sinoatrial portion distends, the common cardial veins disappear and veins originating from the superior and inferior cardinal veins, the terminal portion of vena cava superior and inferior, open directly into the right atrium. The septum spurium and the left venous valves disappear. The right venous valves (superior portion) transform into the crista terminalis of the right atrium, into the valve of the inferior vena cava, and into the valve of the coronary sinus (inferior portion of the right venous valve).

The division of the atrial portion of the heart loop is accomplished by the sagitally oriented septum primum and septum secundum. The septum secundum is sickleshaped. The opening between its free edge and the atrioventricular canal is the foramen primum. As the septum primum reaches the endocardial cushions of the atrioventricular opening, the foramen primum obliterates. The septum primum becomes perforated before the foramen primum obliterates. The newly formed hole, the communication between the atria, is the foramen secundum.

The second sickle-shaped atrial septum secundum, the limbus of the foramen ovale, appears on the right side of the septum primum. Its edge never

52

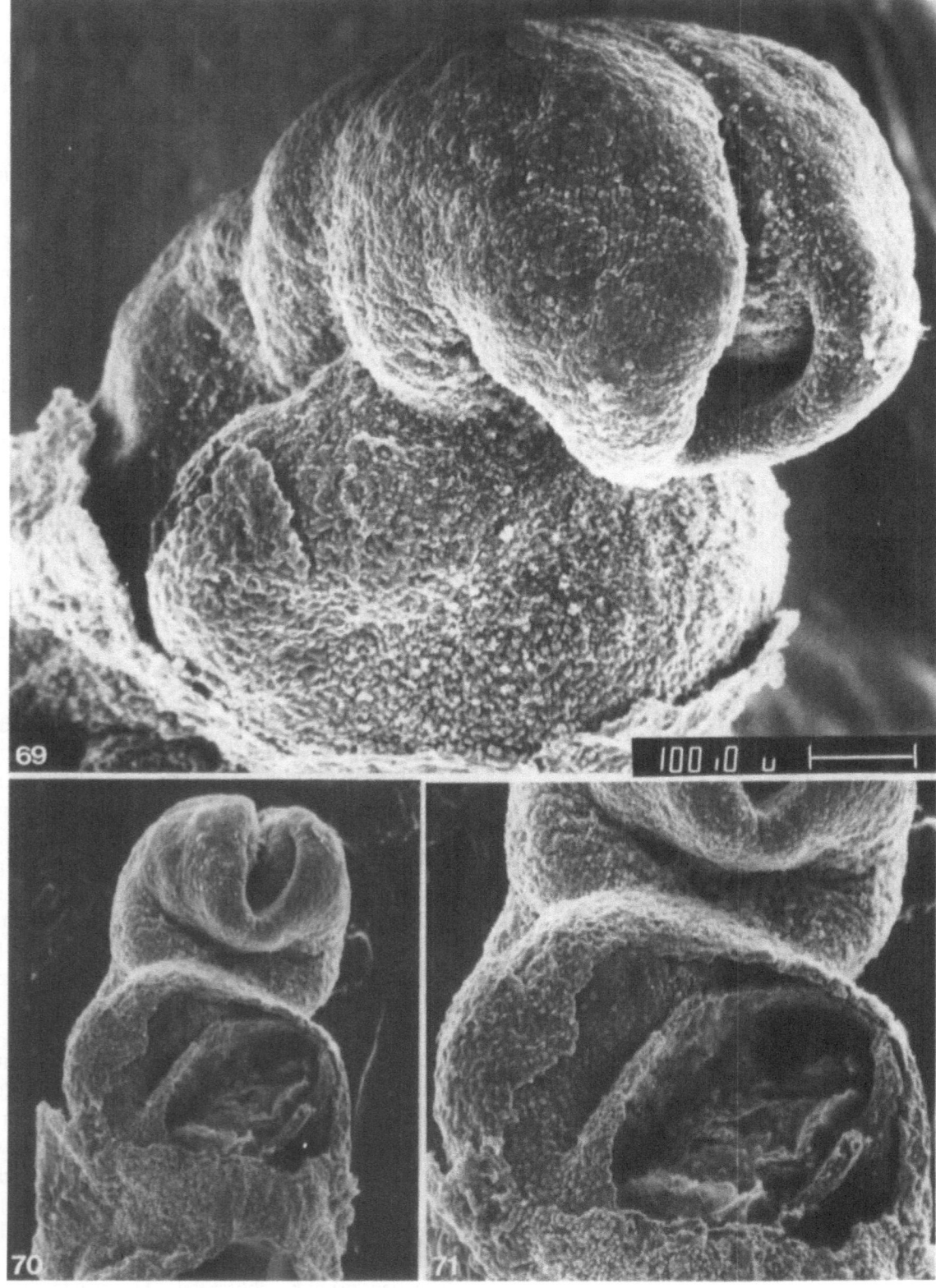

Figures 69–71. (69) Heart prominence (underneath the head with an open anterior neuropore) in a 13-somite embryo. Stage 6-2.
(70) Dissected pericardial cavity with a dissected S-shaped cardiac loop; 13-somite embryo. Stage 6-2. **(71)** Dissected heart loop with exposed myoepicardial mantle and endothelial endocardium; 13-somite embryo. Stage 6-2.

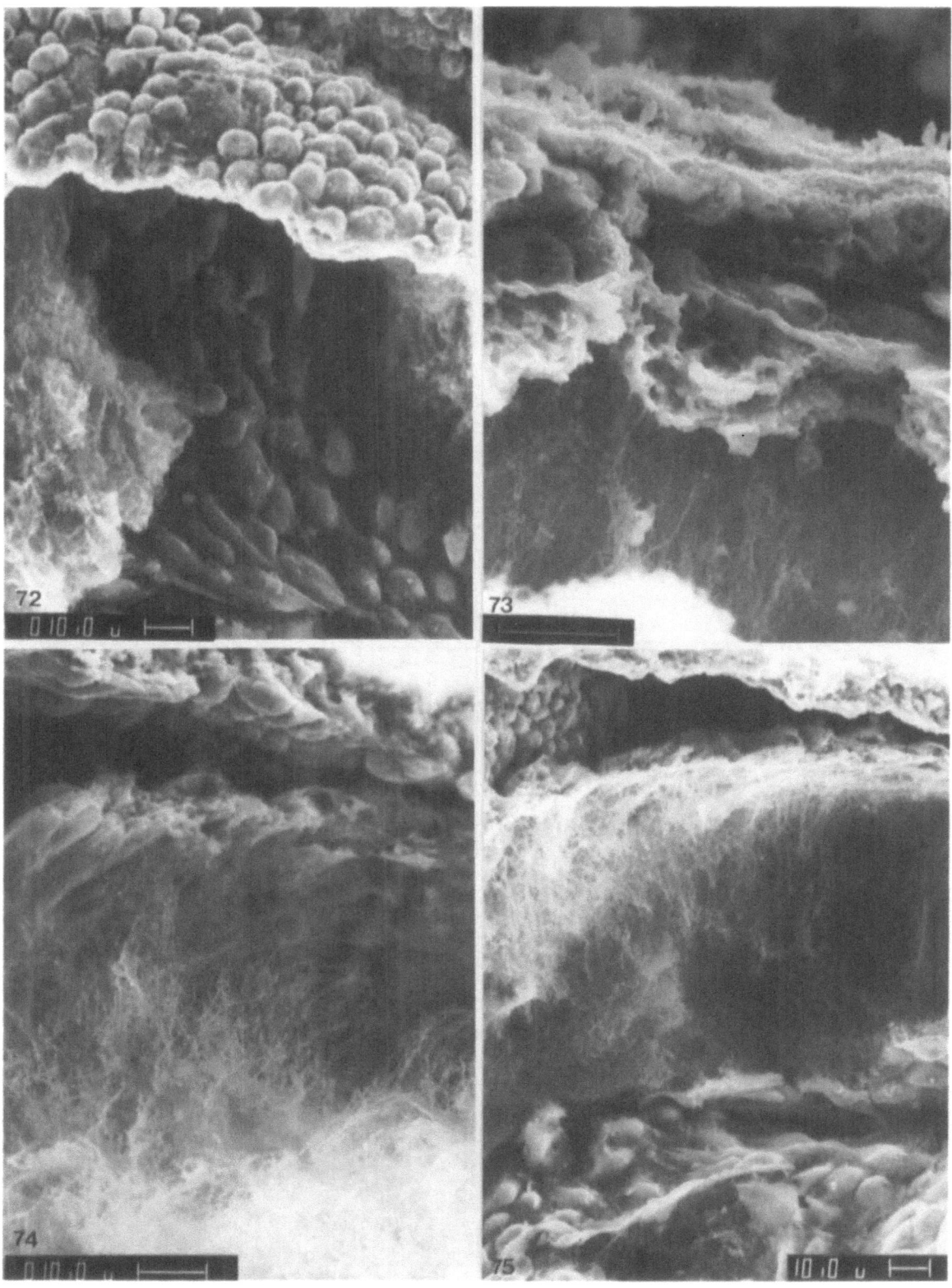

Figures 72–75. Components of the heart loop. Stage 6-2. **(72)** Cells of myoepicardial mantle (top), cardiac jelly (left) and endothelial cells of endocardium (right). **(73)** Epicardium, myoepicardium cells and fibers of cardiac jelly. **(74)** Fibers of the cardiac jelly located between the myoepicardial mantle and the endocardium. **(75)** Components of the heart tube: myoepicardial mantle, cardiac jelly, and endothelial endocardium.

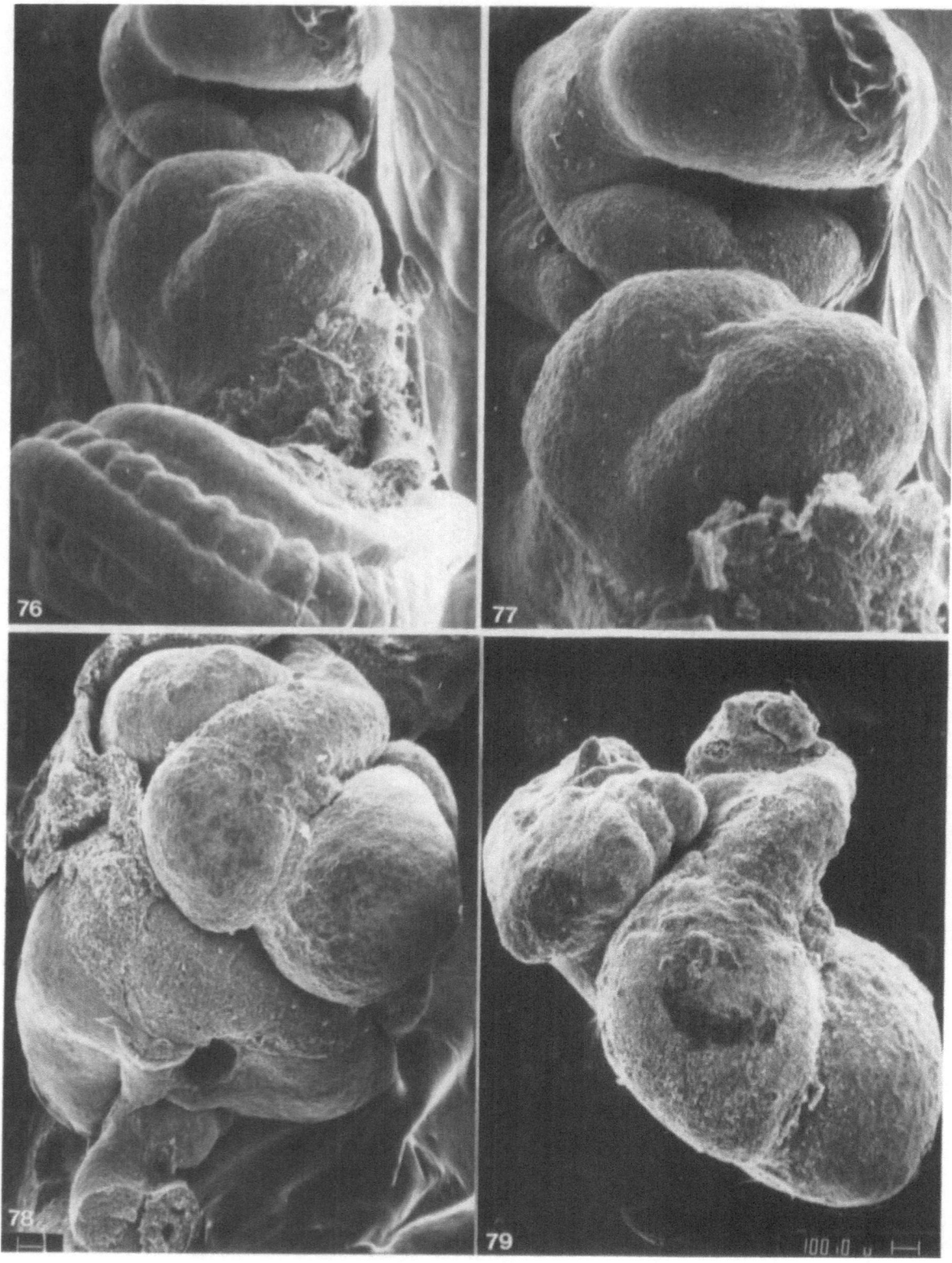

Figures 76–79. (76) S-shaped heart tube located within the pericardial cavity in an embryo. Stage 6-3. **(77)** S-shaped heart tube within the pericardial cavity bulging anteriorly to the pharyngeal arches. Stage 6-3. **(78)** Dissected heart and liver. Stage 7-3. **(79)** The heart seen from the right. Stage 7-3. Structures depicted: right atrium (left, top), right ventricle, conus cordis and truncus arteriosus, interventricular sulcus, and left ventricle.

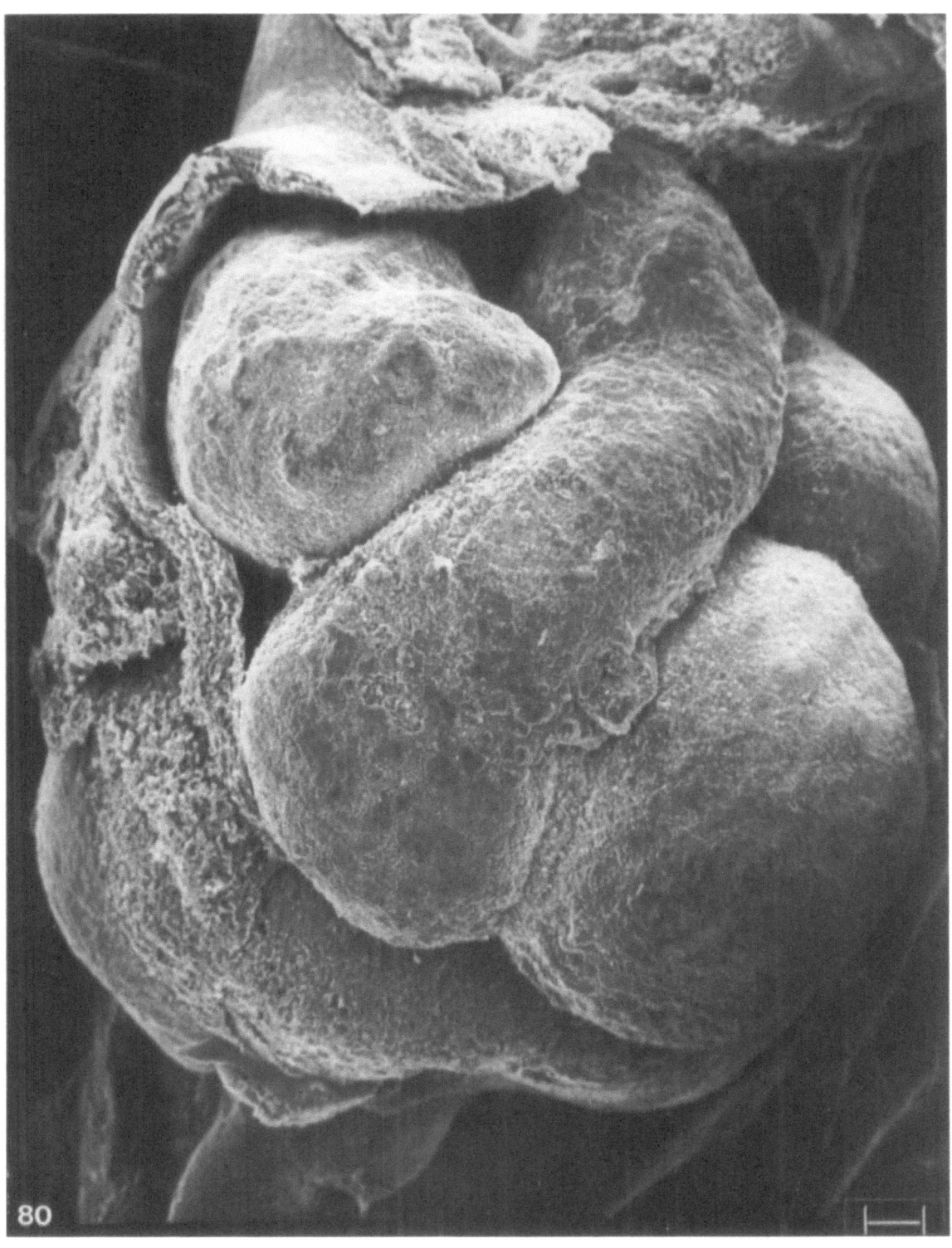

Figure 80. The heart. Stage 7-3. Right oblique view. The pericardium dissected. Structures depicted: right atrium, left ventricle, conus and arterial truncus, interventricular sulcus, left atrium, and left ventricle.

56

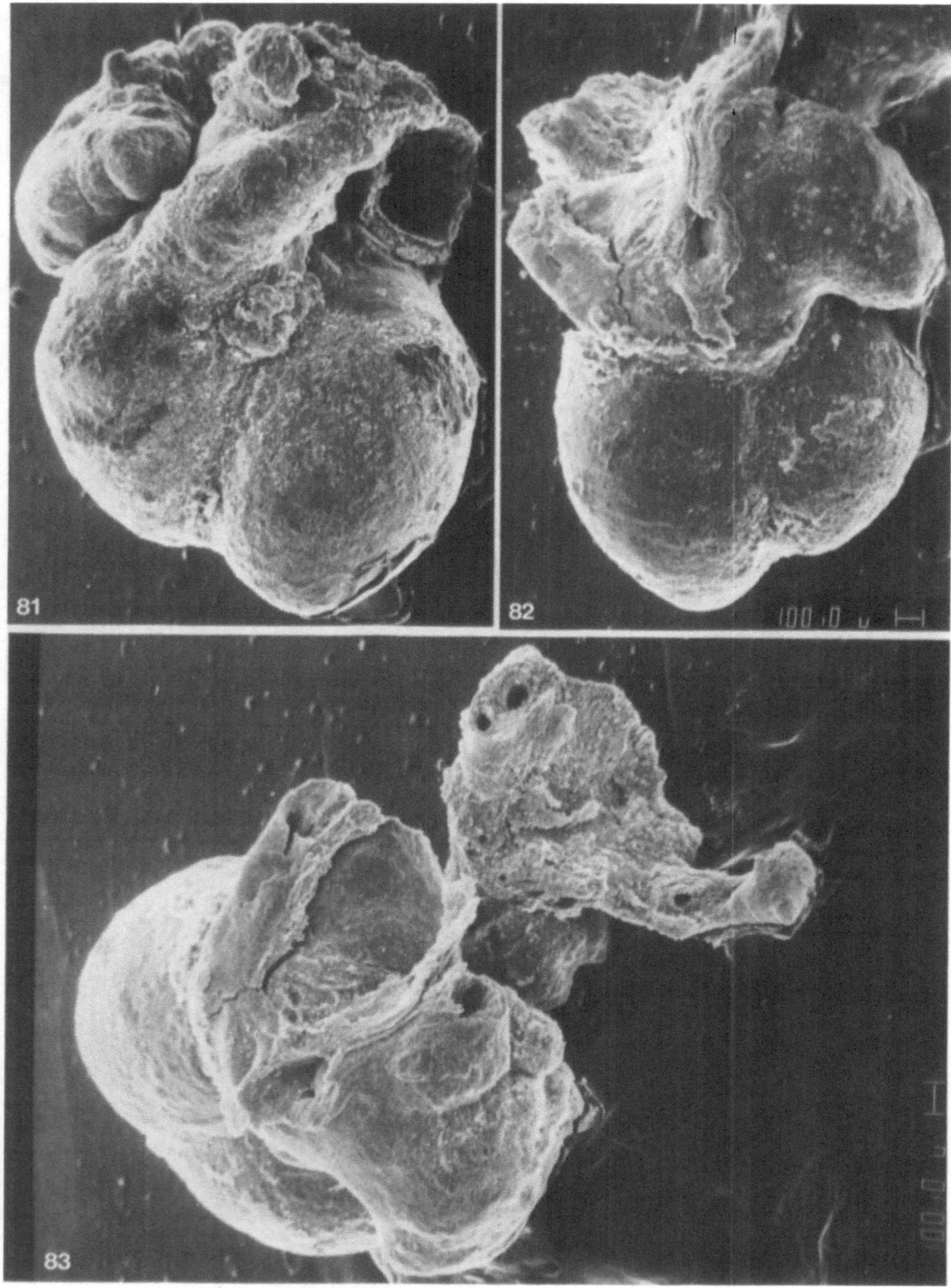

Figures 81–83. (81) The heart. Stage 7-4. Frontal anterior view. The left atrium opened. **(82)** The heart. Stage 7-4. Posterior view. **(83)** Dissected heart and aortic arches. Stage 7-4. Superoposterior view.

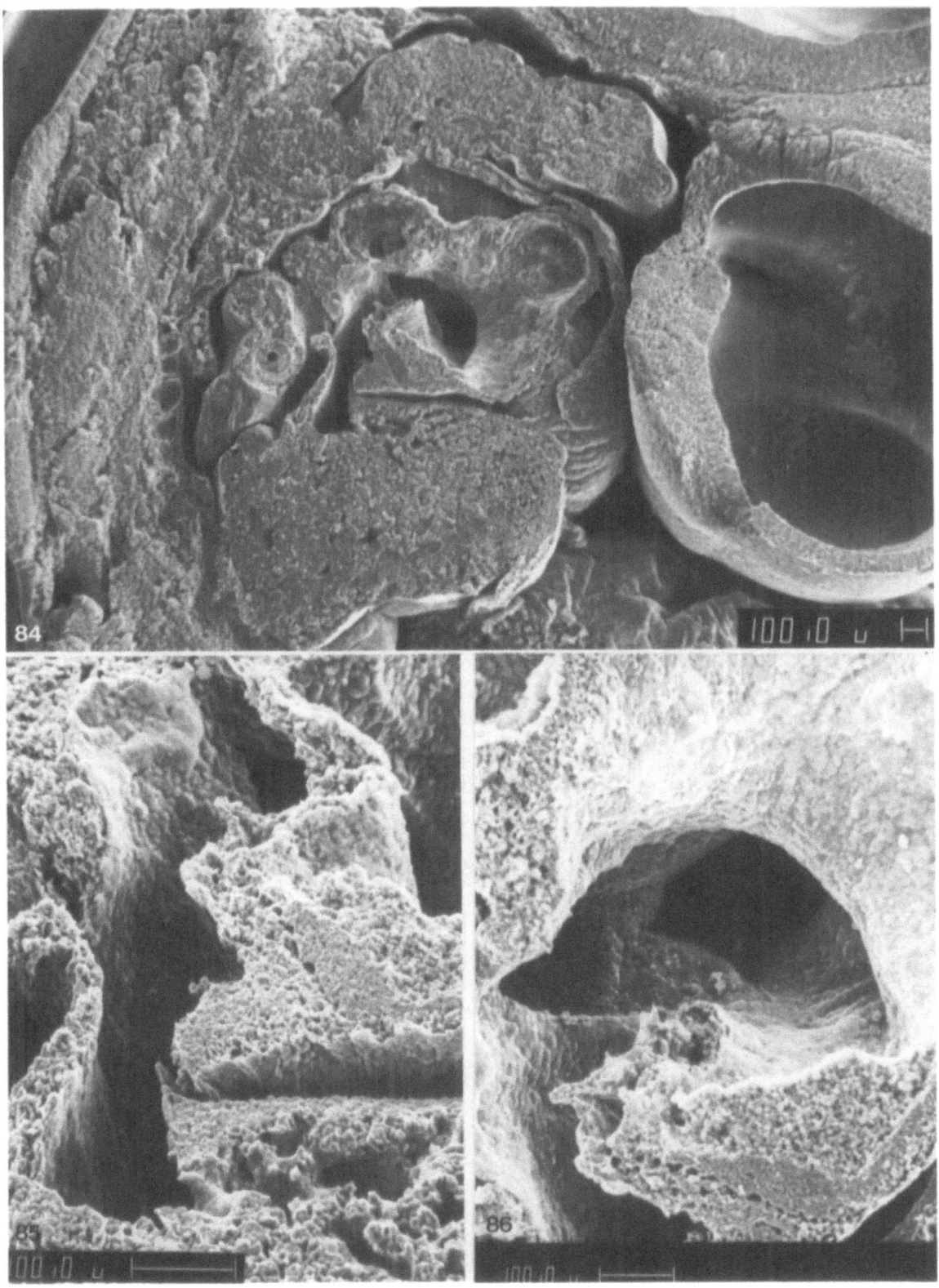

Figures 84–86. (84) Sagittal midline dissection of an embryo. Stage 7-4. Structures depicted: mandible with primordium of the tongue, oral and pharyngeal cavities, dissected pericardial cavity with dissected heart; longitudinally dissected vena cava entering the right atrium, liver and septum transversum, pleural cavity with lungs, closed pericardioperitoneal canal, and remnants of mesonephric malpighian corpuscles posterior to the pleural cavity. **(85)** Longitudinally dissected inferior vena cava. Stage 7.4. **(86)** Opening of the inferior vena cava into the right atrium delineated by the valve, and the right atrioventricular opening. Stage 7-4.

58

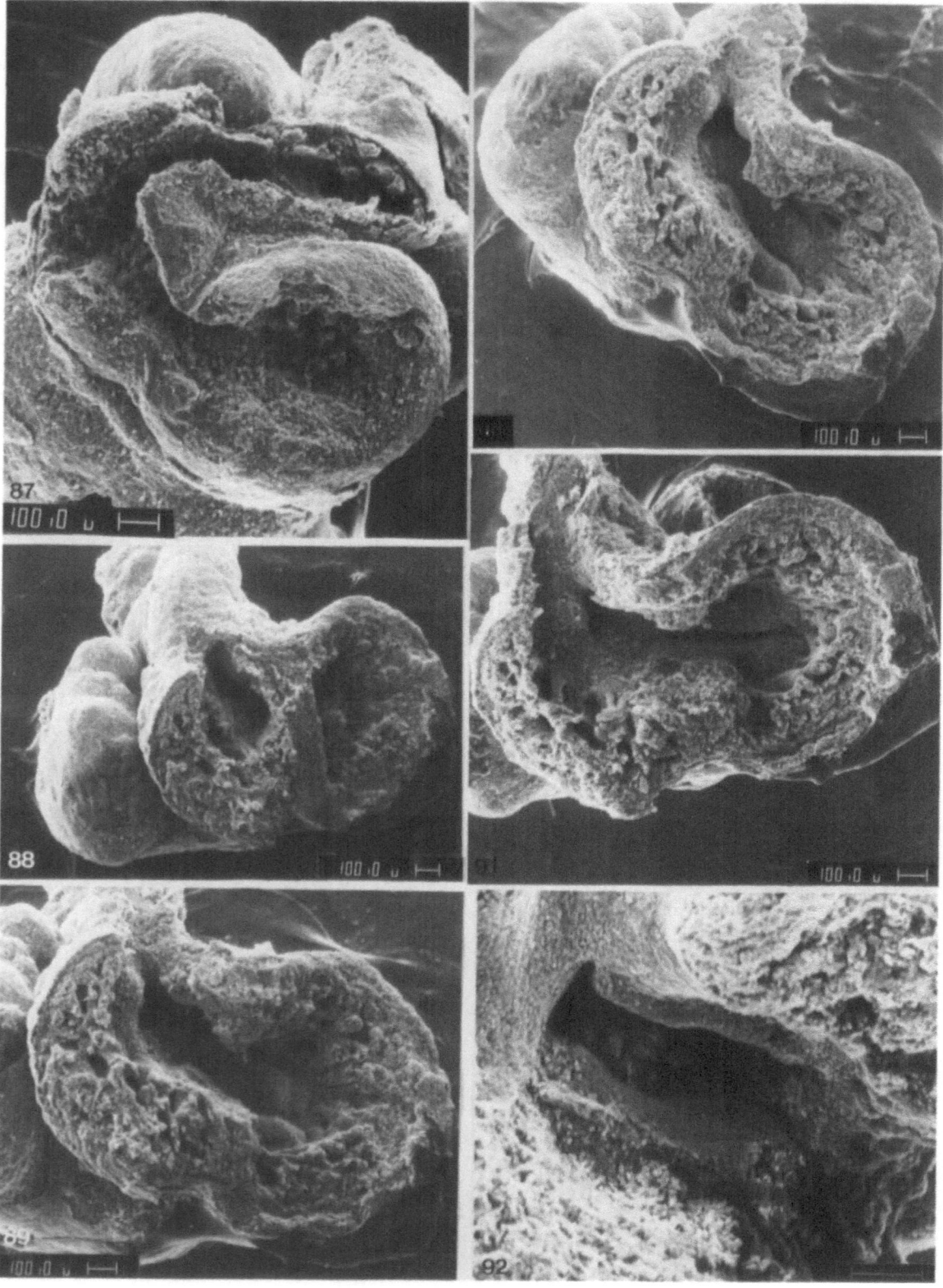

Figures 87–92. Dissected heart. Stage 7-3. **(87)** Dissection of the left ventricle, right ventricle, and arterial conus. **(88)** Dissected left and right ventricles incompletely separated by the interventricular septum. **(89)** Exposed interventricular foramen (the interventricular septum has been removed). Right superior and left inferior truncus swellings are evident. **(90)** Arterial truncus with truncus swellings and a slitlike common atrioventricular opening. **(91)** Opening of the arterial truncus and common atrioventricular opening delineated by superior and inferior endocardial cushions. **(92)** Common atrioventricular canal.

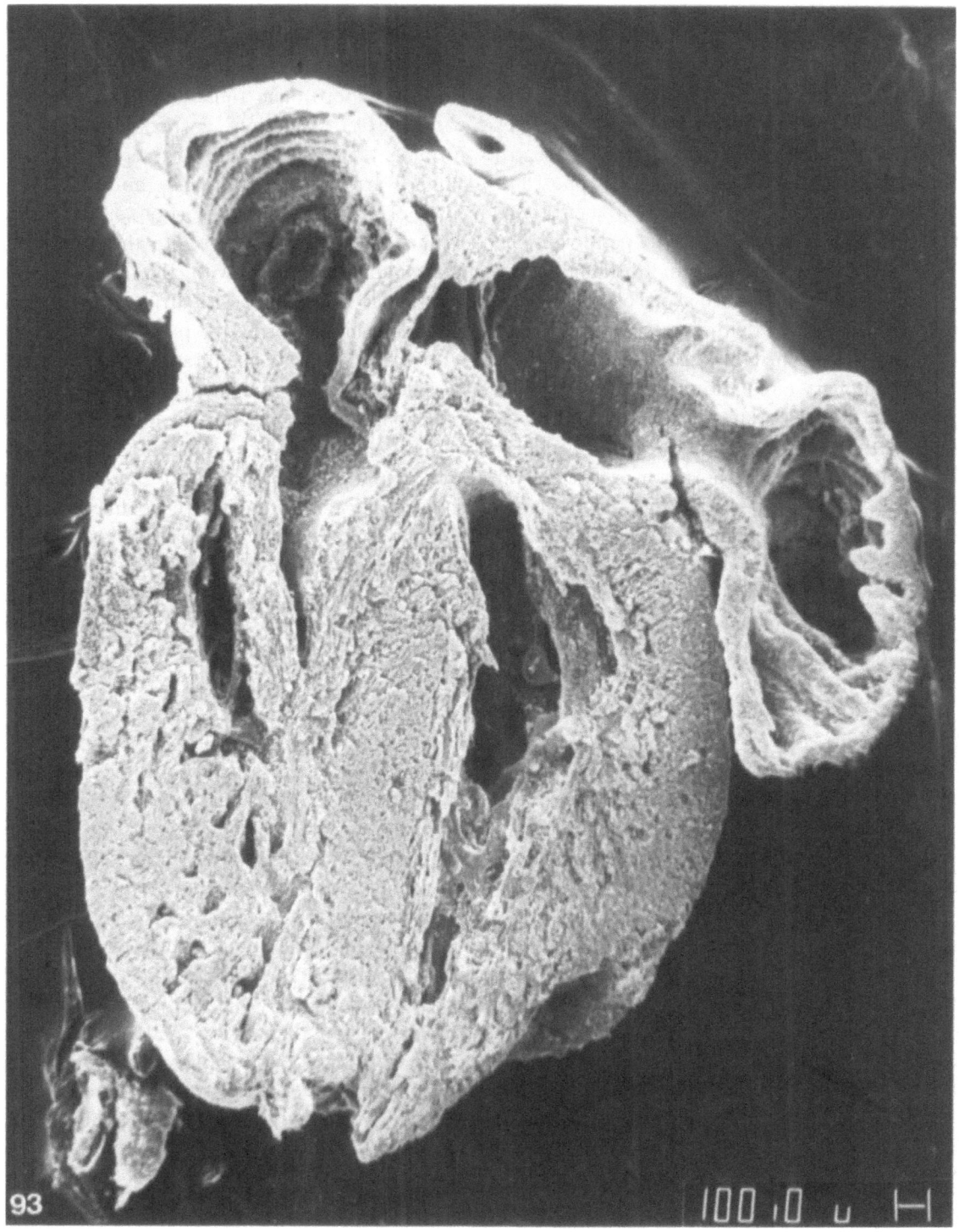

Figure 93. Frontal dissection of the heart. Stage 8-2. Structures depicted: right atrium, right venous valve, interseptovalvular space, septum secundum, septum primium, left atrium with left auricle, right atrioventricular opening, right ventricle, muscular interventricular septum, and left ventricle.

fuses with the endothelial cushions of the atrioventricular canal. The area between the septum secundum and the septum primum and the valves of the venous sinus is known as the interseptovalvular space. The septum secundum is apposed to the septum primum and covers the foramen secundum as a valve enabling blood to pass from the right atrium to the left atrium. Postnatally, the foramen ovale becomes closed as the blood pressure in the left atrium exceeds the pressure in the right atrium in consequence of the tremendously increased pulmonary circulation.

b) The atrioventricular canal of the loop is delineated by two endocardial atrioventricular cushions: an anterior and a posterior. As the central portions of the cushions fuse together, the right and left atrioventricular openings become separated. The central portion of the atrioventricular cushions fuses also with the atrial septum primum. An additional, lateral endocardial cushion appears in the right atrioventricular opening. The three cushions on the right side transform into the tricuspid valve, the two cushions on the left into the bicuspid (mitral) valve.

The cushions are endocardial duplications containing cardiac jelly. The spongy musculature is attached to the jelly. As the heart grows, new muscle layers are apposed to the surface as a myocardial compact layer. The myocardial spongy layer transforms into trabecular and papillary muscles attached to the valves by chordae tendinae.

c) The interventricular septum consists of two portions: a muscular and a membranous. The muscular septum is formed by the bulboventricular fold (superoposteriorly) and by the proper interventricular septum (inferiorly). The position of the interventricular septum is evident on the heart surface as the interventricular sulcus. As the interventricular muscular septum is formed by apposition of new musculature from outside, the interventricular foramen, located in the anterior portion of the septum underneath the atrioventricular openings, remains preserved for a considerable time. The interventricular foramen closes by connective tissue from the inferior endocardial cushion of the heart conus (the membranous portion of the interventricular septum).

d) The aorticopulmonary septum originates from two twisting endocardial ridges protruding into the arterial truncus (the aorticopulmonary segment of the heart loop). The right (superior) ridge fuses with the left (inferior) aorticopulmonary ridge forming the aorticopulmonary septum. The pulmonary artery leaves the heart anteriorly to the aorta. The most inferior portion of the aorticopulmonary septum extends into the conus of the heart tube and fuses with the muscular interventricular septum and with atrioventricular endocardial cushions closing the primary interventricular foramen as the membranous portion of the interventricular septum.

e) Semilunar valves of the aorta and the pulmonary artery originate from endocardial thickenings present in the proximal portion of the arterial truncus. As the aorta and the pulmonary artery become separated, three endocardial tubercles become evident in each vessel. The tubercles are gradually hollowed by the backflow of the blood during ventricular diastoles and transform into the semilunar vela of the valves.

The primitive embryonic circulation

The first embryonic vessels may be classified as intraembryonic (in the embryonic body) and extraembryonic. Extraembryonic vessels are vitelline and chorionic (placental). The heart tube begins to contract at the stage of 10–13 somites and primarily only fluid is pushed into the vascular primordia. The first primitive circulation is the yolk-sac (vitelline) circulation. The fluid is pumped by the heart loop into the primitive aortae and through paired omphalomesenteric (vitelline) arteries into the capillary net of the yolk sac. Blood cells are released from the blood islands of the yolk sac and transported by the way of two omphalomesenteric (vitelline) veins to the heart. The vitelline circulation is the first source of blood cells in early embryos. Parallel to the vitelline circulation, the umbilical, or embryochorionic circulation becomes functional. This circulation supplies embryonic blood to the chorion by the way of two umbilical arteries located in the connecting stalk and connect the caudal aorta with the capillary nets of chorionic villi. The blood coming from the chorion contains oxygen and nutritive substances and returns to the heart by the way of two umbilical veins. After the development of liver, the right umbilical vein undergoes regression. The two umbilical veins are reduced to the single one (left).

The arteries (Fig. 94)

Blood emitted by the embryonic heart comes into the arterial truncus, which splits into two ventral aortae, and then passes through aortic arches into two dorsal aortae that fuse into a single aorta in the thoracic area. Six paired aortic arches are formed subsequently in relation to the pharyngeal arches. The main arteries of the head and proximal extremities are formed in the following way: The first aortic arches disappear. The second aortic arches disappear except for a stapedial artery temporarily supplying the middle-ear cavities (stapedial crura are formed around this artery). The third aortic arches contribute to the proximal portion of the internal carotid arteries. The ventral aortae cranially to the third aortic arches become external carotid arteries.

Segments of both ventral aortae located between the third and fourth aortic arches become common carotid arteries. The segment of the right ventral aorta located caudally to the fourth arch gives rise to the brachiocephalic trunk. The right fourth aortic arch becomes the proximal portion of the right subclavian artery. The left ventral aorta caudal to the fourth arch transforms into the definitive ascending aorta, the fourth aortic arch into the arch of the definitive aorta, and the left dorsal aorta (caudally to fourth arch) changes into the proximal portion of the descending aorta. The sixth aortic arches (located between the arterial truncus and the dorsal aortae) are transformed into pulmonary arteries. The dorsal segment of the left sixth aortic arch persists until birth and is known as the arterial duct of Botallus.

Arterial branches growing from the descending aorta are intersegmental, supplying the somites and the medullary tube, and visceral, destined for the abdominal organs. The vertebral arteries originate from longitudinal anastomoses of intersegmental arteries along the cervical portion of the neural tube. Capillary nets of the proximal limb join the seventh intersegmental artery and contribute to subclavian arteries. The right seventh intersegmental artery joins the fourth aortic arch. Typical segmental branches are the intercostal and lumbar arteries. The main arteries of the distal limb develop from primitive umbilical arteries that are originally

the main terminal visceral branches of the aorta. As the roots of the primitive umbilical arteries become replaced by a secondary connection to the more caudally located parietal aortic branches, the newly formed connection transforms into the common iliac artery. Each common iliac artery branches into the internal and external iliac arteries. Umbilical arteries lose their origin from the aorta and become branches of the internal iliac arteries.

Visceral aortic branches may be classified as lateral – destined for retroperitoneal organs related to the mesonephric ridges (such as arteriae phrenicae abd., arterieae suprarenales, arterieae renales, and arterieae testiculares or ovaricae), and as ventral – for the derivatives of the digestive tube. Paired primordia of the ventral visceral arteries fuse secondarily into single arteries. The coeliac artery originates from two ventral intersegmental arteries: the cranial mesenteric artery from 11th to 14th ventral intersegmental arteries, which are related to the vitelline arteries, and the caudal mesenteric artery from the 20th ventral intersegmental arteries.

The veins

The cardinal veins. The blood from the head of the early embryo is drained by the paired anterior cardinal veins, and from the body by the posterior cardinal veins. The anterior and posterior cardinal veins join and as the common cardinal veins (ducts of Cuvier) enter the venous sinus of the heart tube laterally (in addition to the umbilical veins and to the medially located vitelline veins). As the anterior limbs develop and subclavian veins are formed, a large brachiocephalic anastomosis is formed between the anterior cardinal arteries (at the level of the openings of the subclavian veins). Consequently the proximal portion of the left anterior cardinal vein obliterates.

The superior vena cava is formed by the right common cardinal vein and by a segment of the right anterior cardinal vein, which is located below the level of the brachiocephalic anastomosis. The brachiocephalic anastomosis transforms into the left brachiocephalic (innominate) vein. The segment of the right anterior cardinal vein between openings of the left brachiocephalic vein and the right subclavian vein becomes the right brachiocephalic vein. The upper portion of cardinal veins (anterior to the openings of brachiocephalic veins)

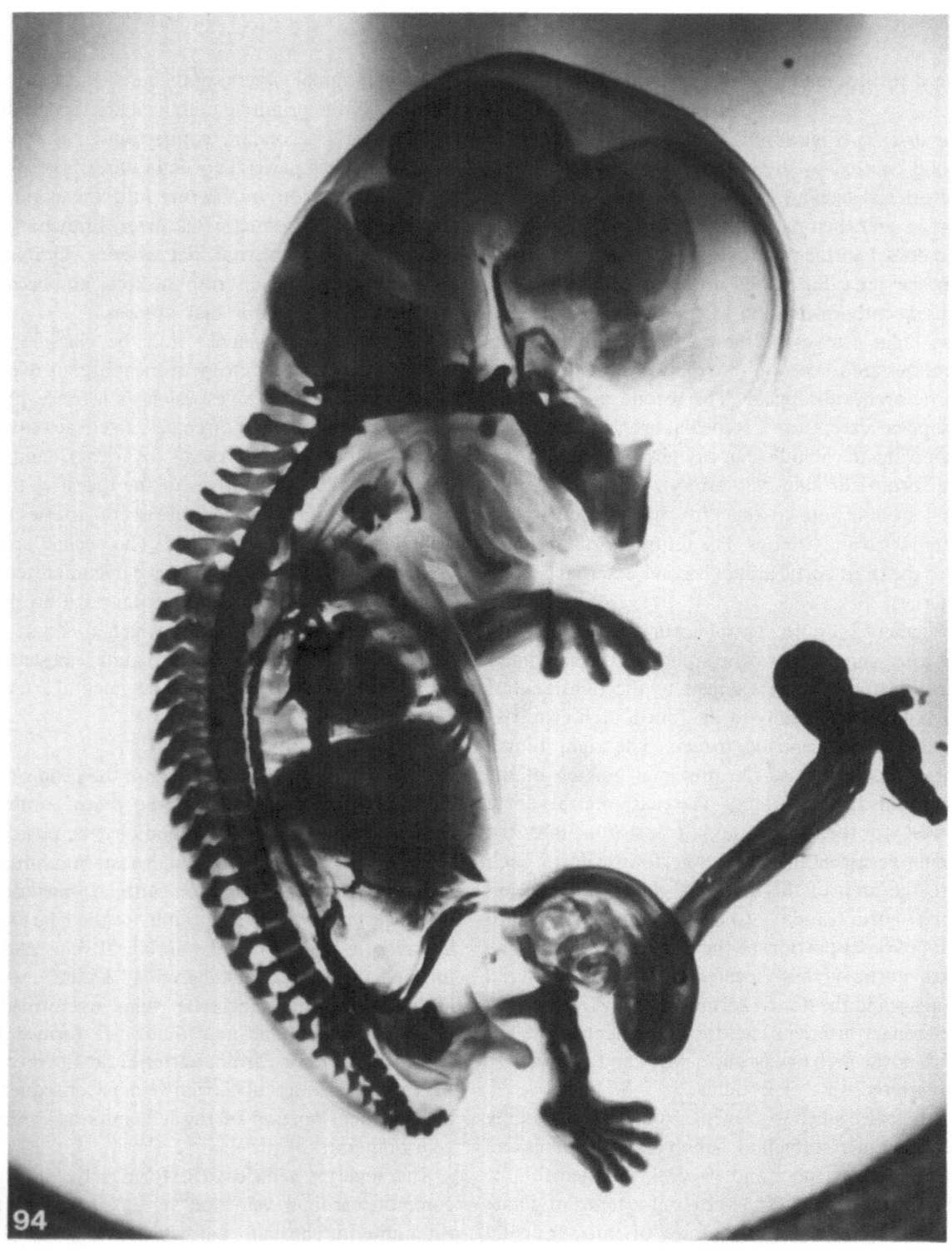

Figure 94. Cleared embryo with visualized main vessels. Stage 8-2. Structures depicted: umbilical vein, inferior vena cava, ductus arteriosus, aortic arcus and descending aorta, and umbilical arteries.

turns into the internal jugular veins. External jugular veins are secondary channels draining facial capillary plexuses.

The inferior vena cava emerges from a complicated development involving systems of the posterior cardinal veins, the infracardinal veins, and the supracardinal veins. The infracardinal veins collect blood from mesonephros. The supracardinal veins are formed dorsal to the posterior cardinal veins. Infra- and supracardinal veins open within the thoracic area into the posterior cardinal veins. In the low lumbar area, anastomoses are formed between the infra- and supracardinal veins. A big interiliac anastomoses originates in relation to the development of iliac veins, which drain the posterior limbs. In the retrohepatic area, a big intersubcardinal anastomosis connects subcardinal veins of both sides. In relation to the asymmetrical development of the liver and heart, and to a complicated hemodynamic relationship in the abdominal cavity, blood from the abdominal organs and lower limbs returns to the heart by various shunts. The following four segments are distinguished on the definitive inferior vena cava:

1) The terminal segment (between the liver and the right atrium), contributed to by the right vitelline vein.
2) The hepatic segment, formed from the hepatosubcardinal anastomosis between the right vitelline vein and the right subcardinal vein.
3) The renal and subcardinal segments, formed from the right subcardinal vein and the left renal vein. (The left renal vein is derived from the intersubcardinal anastomosis, which receives the left suprarenal, left renal, and left gonadal veins. Corresponding veins from the right side open directly into the renal segment of the inferior vena cava.)
4) The supracardinal segment originating from the anastomosis between the right subcardinal and supracardinal veins and from the right supracardinal vein, which receives the right common iliac and iliolumbar veins. The left common iliac vein originates from the interiliac anastomosis.

The vitelline (omphalomesenteric) veins. Vitelline veins drain the capillary plexus and blood islands of the yolk sac and enter the venous sinus of the embryonic heart medially. The liver cords grow into the vitelline vein and change them into the network of hepatic sinusoids. The intraembryonic distal segment of the right vitelline vein and its mesenteric branches change into the portal vein. The proximal portion of the right vitelline vein contributes to the terminal segment of the inferior vena cava. From the left vitelline vein, only a portion transforming into the left hepatic vein remains preserved.

The umbilical veins. There are originally two umbilical veins lateral to the allantois within the connecting stalk. They bring blood from the chorionic vessels to the heart. The right umbilical vein obliterates. The left anastomosis with the hepatic veins and a large canal, the venous duct of Arantius, is formed on the ventral side of liver. The venous duct opens into the inferior vena cava.

Hematopoesis
The first hematopoetic tissues are the blood islands formed by mesenchymal cells in the wall of the yolk sac. During the eighth week, yolk-sac blood islands become exhausted and hematopoetic tissue is present in the liver. At the same time, the first lymphatic cells can be identified in the thymic primordium, in the cervical and retroperitoneal lymph nodes, and in the spleen. At 10–12 weeks, hematopoetic tissue appears in the bone marrow. Hematopoesis in the liver and spleen diminishes during the second trimester and disappears in the perinatal period.

Fetoplacental circulation
Beginning with the second month postconceptionally, the development of the embryo depends on the fetoplacental circulation. Oxygenated blood from chorionic villi comes to the fetus by the umbilical vein located within the umbilical cord. The umbilical vein bypasses the liver as the venous duct and sends a branch connecting with the portal hepatic circulation. The venous duct (with oxygenated blood) accepts hepatic veins (with venous blood) and enters the right atrium of the heart, which accepts venous blood also from the superior vena cava. From the right atrium, most of the blood goes through the foramen ovale into the left atrium, bypassing the right ventricle, and from the left atrium into the left ventricle and into the aorta. A

much smaller portion of blood enters the right ventricle and is divided into two portions: one of them enters pulmonary nutritive circulation and the other passes directly through the arterial duct (of Botallus) from the pulmonary artery into the aorta. The blood from the aorta is distributed within the fetal body. The blood from the common iliac arteries passes into the two main umbilical arteries, which enter the umbilical cord and bring blood to the chorionic villi.

Within fetal circulation, maximum oxygenated blood is only in the umbilical vein and in the venous duct before openings of the hepatic veins. The maximum deoxygenated blood is in the superior vena cava and in the inferior vena cava before its opening into the venous duct, and in the pulmonary veins and in the portal vein before accepting the communicating branch from the venous duct. After delivery. the lungs distend with the inhaled air. The blood pressure rises in the left atrium of the heart and diminishes in the right atrium. Consequently the foramen ovale closes and the arterial duct of Botallus obliterates. The umbilical vessels obliterate after the placenta has been removed. The venous duct regresses, tranforming into a connective tissue band–ligamentum venosum of the liver.

The lymphatics
Lymphatics represent channels returning tissue fluid to the venous blood. Lymphatic vessels originate from mesenchyme. Two jugular lymph sacs are present in the neck at the end of second month. These sacs are connected with the lymphatics of the upper limbs and with the retroperitoneal lymph sac draining the pleuropericardial cavity. The upper portion of this sac changes into the thoracic duct, the lower portion into the cisterna chyli draining the lymphatics of the gastrointestinal tract. The thoracic duct opens into the superior vena cava near to the opening of the internal jugular vein.

The lymph nodes develop by accumulation of lymphocytes at places of branching lymphatic capillaries.

The thymus has a double origin: The endodermal epithelium is detached from the ventrolateral portion of the third pharyngeal pouch. The thymic reticulum and the Hassal corpuscles are

epithelial derivatives. The capsule, septa, and T-lymphocytes are mesenchymal in origin. (An alternative proposal considers the Hassal corpuscles as derivatives of the ectoderm of the third pharyngeal pouch.)

The spleen originates from mesenchymal proliferation within the dorsal mesogastrium.

The urogenital organs and adrenals

Urinary and genital organs are closely related in development.

The uropoetic organs
Three subsequent developmental stages of uropoetic organs are known as the pronephros, the mesonephros, and the metanephros. All of them originate from intermediate mesoderm. The basic units of all uropoetic organs are the nephrons.

The pronephros. The pronephros of human embryos is rudimentary, consisting of clusters of cells and rudimentary tubules lateral to the somites. Distinct pronephric tubules are present adjacent to segments 9–12 or 13. A longitudinal cellular cord located dorsolaterally from the pronephric nephrons represents the primary ureter (the mesonephric, or Wolffian, duct). Most consider the primary ureter as a mesodermal structure, while others consider it as an ectodermal derivative.

The mesonephros (Figs. 97–101). The intermediate mesoderm, extending between the last cervical to the second lumbar somite (segments 10–25), represents the mesonephric blastema giving rise to the mesonephros. The mesonephric nephrons differentiate craniocaudally and connect with the primary ureter growing from the pronephric area longitudinally along the mesonephric blastema. The primary ureter, originally draining the pronephros, is known as the mesonephric (Wolffian) duct. The mesonephric nephrons bulge longitudinally along the dorsal attachment of the mesentery into the coelomic (peritoneal) cavity as the urogenital ridges. The urogenital ridges, which extend from thoracic area into the lumbar area, are covered by a mesodermal somatopleura and are anchored by a

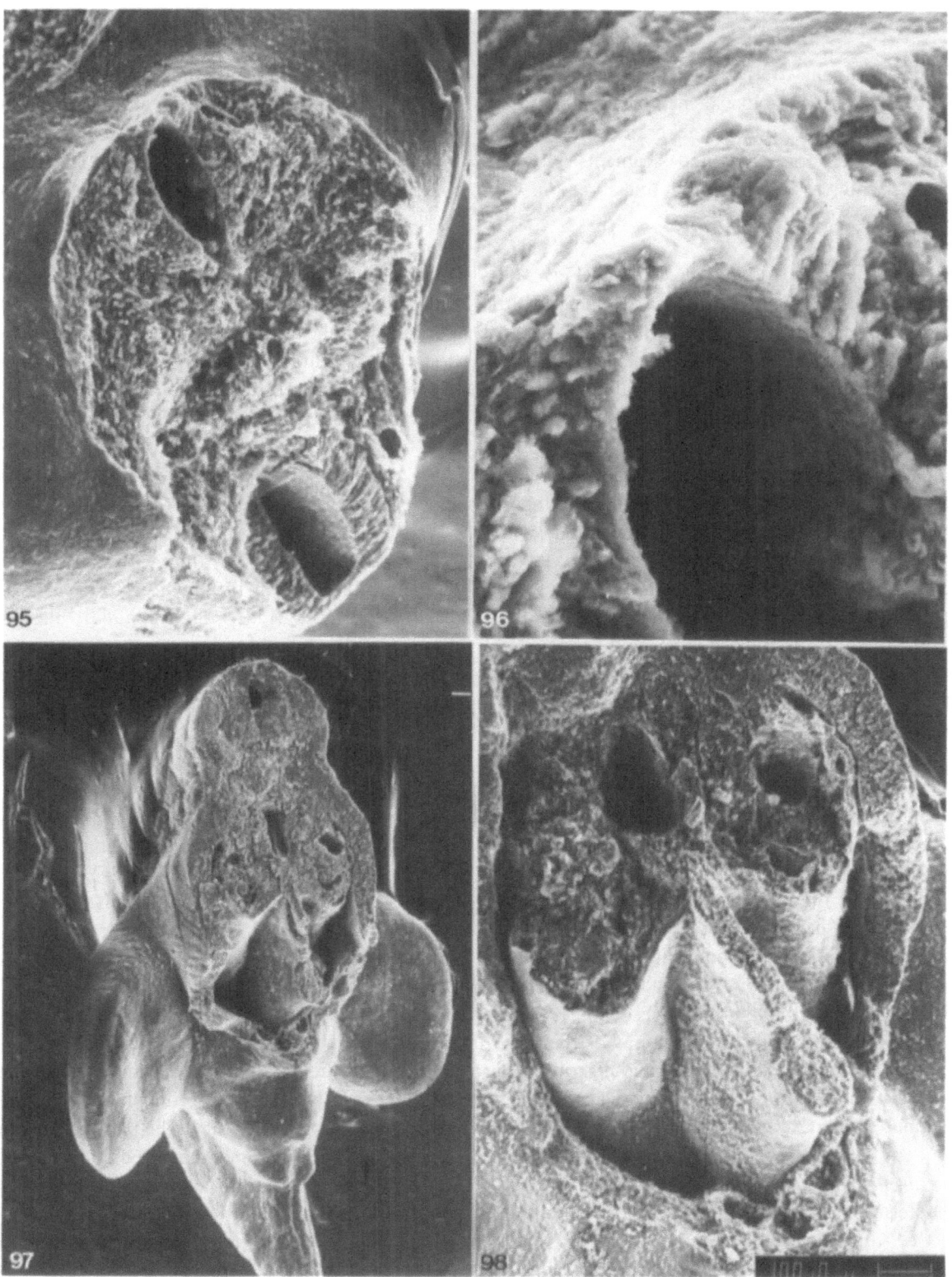

Figures 95–98. (95) Dissected caudal end of an embryo. Stage 7-2. Transversely dissected cloaca (anteriorly) and medullary tube (posteriorly). **(96)** Detail of dissected cloaca closed (anteriorly) by the cloacal membrane. **(97)** Lower end of a transversely dissected embryo. Stage 7-3. Structures depicted: urogenital ridges, mesentery with gut, posterior limb buds, and primordia of indifferent external genitalia. **(98)** Dissected urogenital ridges, and mesentery with the gut, protruding into the peritoneal cavity. Stage 7-3.

caudal ligament to the connective tissue located dorsolateral to the cloacal membrane. The mesonephric nephrons degenerate craniocaudally and only mesonephric epigenital segments close to gonads and the mesonephric ducts remain preserved in males.

The Metanephros (definitive kidney) (Figs. 46c–48c). The kidneys develop from the metanephric blastema derived from intermediate mesoderm of segments 27 and 28 (L-4 and L-5). The metanephric blastema is dorsolateral to the caudal portion of the urogenital ridge. Differentiation of the metanephric blastema into the nephrons of the definitive kidney depends on the ingrowth of the ureteral bud from the mesonephric duct. Derivatives of the ureteral bud give rise to the entire collecting system of the kidney: to the collecting tubules, renal calyces, and renal pelves. The kidneys are retroperitoneal to and 'ascend' from lumbar segments 4 and 5 to thoracic segment 12 by disconcordant growth of the fetal trunk and the kidney.

The ureters
The ureters are derivatives of the mesonephric ducts. The mesonephric ducts open into the anterior portion of the hindgut, which becomes the cloaca. The ureteric buds evaginate before these openings and grow dorsally into the metanephric blastema. As the terminal portions of both mesonephric ducts become incorporated into the endodermal vesicoureteral primordium (derived from the anterior portion of the cloaca), the openings of the ureteral bunds and mesonephric ducts separate. Consequently the openings of the mesonephric ducts are disposed caudally into the ureteral primordium and the vesical trigonum becomes evident.

The urinary bladder (Figs. 95, 96 and 147–150)
The terminal portion of the gut (the cloaca) is divided during weeks 6 and 7 by a transverse peritoneal fold, the urorectal septum, into a dorsally located anorectal canal and a ventrally located primitive urogenital sinus. As the primary ureters (Wolffian ducts) enter the primitive urogenital sinus, the portion of urogenital sinus located above their openings is known as the vesicourethral primordium. The allantois extends from the vesico-urethral primordium into the umbilical cord. The segment of the primitive urogenital sinus located below the openings of the mesonephric ducts represents the definitive urogenital sinus. The urinary bladder originates from the upper portion of the vesicourethral primordium, except for the mesodermal trigonum and a small area of the apex originating from allantois. Most of allantois obliterates and changes into the fibrous urachus (the median umbilical ligament). The lower portion of the vesicourethral primordium contributes to the entire female urethra or to the intramural and upper prostatic portions of the male urethra.

The genital organs (Figs. 49c–74c and 104–146)

The gonads. The testes, or the ovaries, arise from a gonadal blastema on the medioventral surface of each urogenital ridge. The gonadal blastema, known as the genital ridge, consists of primitive mesodermal (mesoblastic, coelomic) cells, mesenchymal cells, and primordial germ cells.

The indifferent genital ducts and external genitalia. After development of the genital ridge, the former urogenital ridges are known as the mesonephric ridges. The genital ridges extend from Th-6 to S-1 in males and from Th-6 to S-2 in females. Proliferating primitive coelomic cells spread out from the cranial end of the genital ridge to the cranial end of the regressing mesonephric ridge and invaginate there as a cellular cord growing laterally along the mesonephric duct, contributing to the primordium of the paramesonephric duct. The primordium of the paramesonephric duct grows along the lateral surface of the mesonephric duct, which serves as a guiding structure. The paramesonephric (Müllerian) ducts become lumenized craniocaudally and cross mesonephric ducts ventrally before approaching the urogenital sinus. The terminal segments of both parallel paramesonephric ducts fuse into a single uterovaginal primordium. At the indifferent stage of genital duct development, the mesonephric ducts open into, while the tips of the paramesonephric ducts only contact, the primitive endodermal urogenital sinus. The protruding area of contact is known as the paramesonephric (Müllerian) tubercle. The primitive urogenital sinus is separated from the hindgut by the urorectal septum and is closed by the sagittally oriented genital membrane. The anterior portion of the genital membrane extends as the epithelial plate to the tip of the genital tubercle. The labioscrotal swellings bulge lateral to the genital tubercle. The connective tissue of the labioscrotal swellings represents the area of insertion of the caudal ligaments of the urogenital ridges.

In males, genital ridges transform into the embryonic testes. Primordial germ cells become incorporated into testicular cords formed by Sertoli cells of mesoblastic origin. Germ cells located within the testicular cords transform into the spermatogonia, which do not begin meiosis until puberty. Embryonic testes produce a water-soluble substance that diffuses along the caudal ligament of the urogenital ridge (the gubernaculum) and causes regression of the ipsilateral paramesonephric (Müllerian) duct. At the beginning of fetal period, the embryonic testes transform into the fetal testes characterized by the presence of interstitial cells producing the androgens: androstenedione and testosterone. Testosterone and its reduced derivative, dihydrotestosterone, are the principal androgens related to the growth of epigenital mesonephric tubules, mesonephric ducts (which transform into the ductus deferentes and the formation of the male external genitalia and accessory sex glands.

In females, at the end of second month, the genital ridges transform into the embryonic ovaries. The intensive mitotic proliferation of the germ cells located within the embryonic ovaries underneath the surface epithelium was considered in classic embryology as 'the secondary proliferation of sex cords'. At the stage of early fetal ovary, at the end of third month, the first oocytes entering meiotic prophase are present. At the stage of late fetal ovaries, at about week 16 postconceptionally, primary ovarian follicles (characterized by a complete layer of granulosa cells) are evident. Formation of primary ovarian follicles occurs only in the presence of two X chromosomes. Meiotic prophase in oocytes precedes their incorporation into the follicles.

The male genital ducts and external genitalia. Normal development of mesonephric derivatives, the epididymis and the vas deferens, is testosterone dependent. Growth of male accesory sex glands, such as the seminal vesicles, the prostate, and the bulbourethral glands, and male differentiation of the external genitalia are dihydrotestosterone dependent. Testosterone is produced by interstitial cells of the fetal testes. Its reduction to dihydrotestosterone occurs in target tissues, where the physiologically active hormone is bound to a specific cytoplasmic protein receptor. Androgen-stimulated epigenital mesonephric tubules connecting the rete testis with the mesonephric duct transform into the ductuli efferentes of epididymis. The proximal portion of the mesonephric duct, incorporated in the mesonephric ridge, becomes the duct of the epididymis. The distal portion of the mesonephric duct, located between the mesonephric ridge and the urogenital sinus, is known as the vas deferens, and its terminal portion (between the opening of the seminal vesicle and the urethra) is the ejaculatory duct. Seminal vesicles originate as diverticula of the preterminal portion of the mesonephric ducts.

During masculinization of the external genitalia, the labioscrotal swellings fuse anteriorly to the anus. The rhaphe of the scrotum is formed by this fusion, which extends anterior to the urethral groove originating from the urethral plate. As the rims of the urethral groove fuse, the penile urethra is formed. The terminal, glandular, portion of the urethral groove is the epithelial glandular plate. The fossa navicularis develops from the epithelial plate by dehiscence of the inner layers of the epithelium. The male urethra is composed of the following portions: the intramural and upper prostatic (from the vesico-urethral primordium), the lower prostatic and the membranous (from the urogenital sinus), and the cavernous (penile and glandular) from the urethral plate. The prostatic glands and the bulbo-urethral glands are endodermal derivatives of the urogenital sinus.

The female genital ducts and external genitalia. Female genital ducts, the oviducts and the uterus, are derivatives of the paramesonephric (Müllerian) ducts. The superior (mesonephric) portions of paramesonephric ducts remain preserved as the uterine tubes (oviducts). The inferior portions of both paramesonephric ducts fuse into a single uterovaginal primordium with a uterine segment and a vaginal segment. Within the uterine segment, the amount of mesenchyme accompanying paramesonephric ducts as well as the mesenchyme accompanying the mesonephric ducts increases and contributes to the myometrium. The epithelium of the fused paramesonephric ducts provides the uterine lining. The epithelium of mesonephric ducts degenerates. The vaginal segment of the uterovaginal primordium contacts the endodermal epithelium of the paramesonephric tubercle protruding between the openings of mesonephric ducts into the urogenital sinus. The vaginal plate is formed by a multilayered squamous epithelium around the vaginal segment of the uterovaginal primordium, which is lined by cylindrical paramesonephric epithelium. The vagina becomes lumenized by degeneration of the paramesonephric epithelium and by dehiscence of the vaginal plate.

The hymen develops as an incomplete septum between the vagina and the urogenital sinus in the area of the paramesonephric tubercle. The urogenital sinus elongates sagitally during feminization of the external genitalia and during formation of the vaginal plate. The female urogenital sinus contributes to the vaginal vestibule. The female urethra is derived entirely from the lower portion of the vesicourethral primordium.

During feminization of the external genitalia, the indifferent genital tubercle (phallus) transforms into the clitoris. The urethral plate does not close and its rims become the labia minora. The associated cavernous tissue contributes to the bulbus of the vestibulum. The labioscrotal swellings transform into the labia majora.

The adrenals (Figs. 151 and 152)

The cortex of the adrenals is mesodermal; the medulla is neuroectodermal. The cortex is formed by primitive coelomic (mesoblastic) cells ingrowing into the mesenchyme near the root of mesentery medial to the upper portion of the urogenital ridges. The coelomic cells condense into a blastema that differentiates into cords of the fetal adrenal cortex. The primordium of the definitive cortex appears as a distinct subcapsular layer on the surface of fetal cortex during the third fetal month.

The adrenal medulla is contributed to by sympathetic neuroblasts migrating into the fetal cortex from the neural crest. The neuroblasts form multiple clusters within the fetal cortex and change into chromaffin cells. After birth, as the fetal cortex degenerates, the clusters of chromaffin cells fuse into the centrally located medulla.

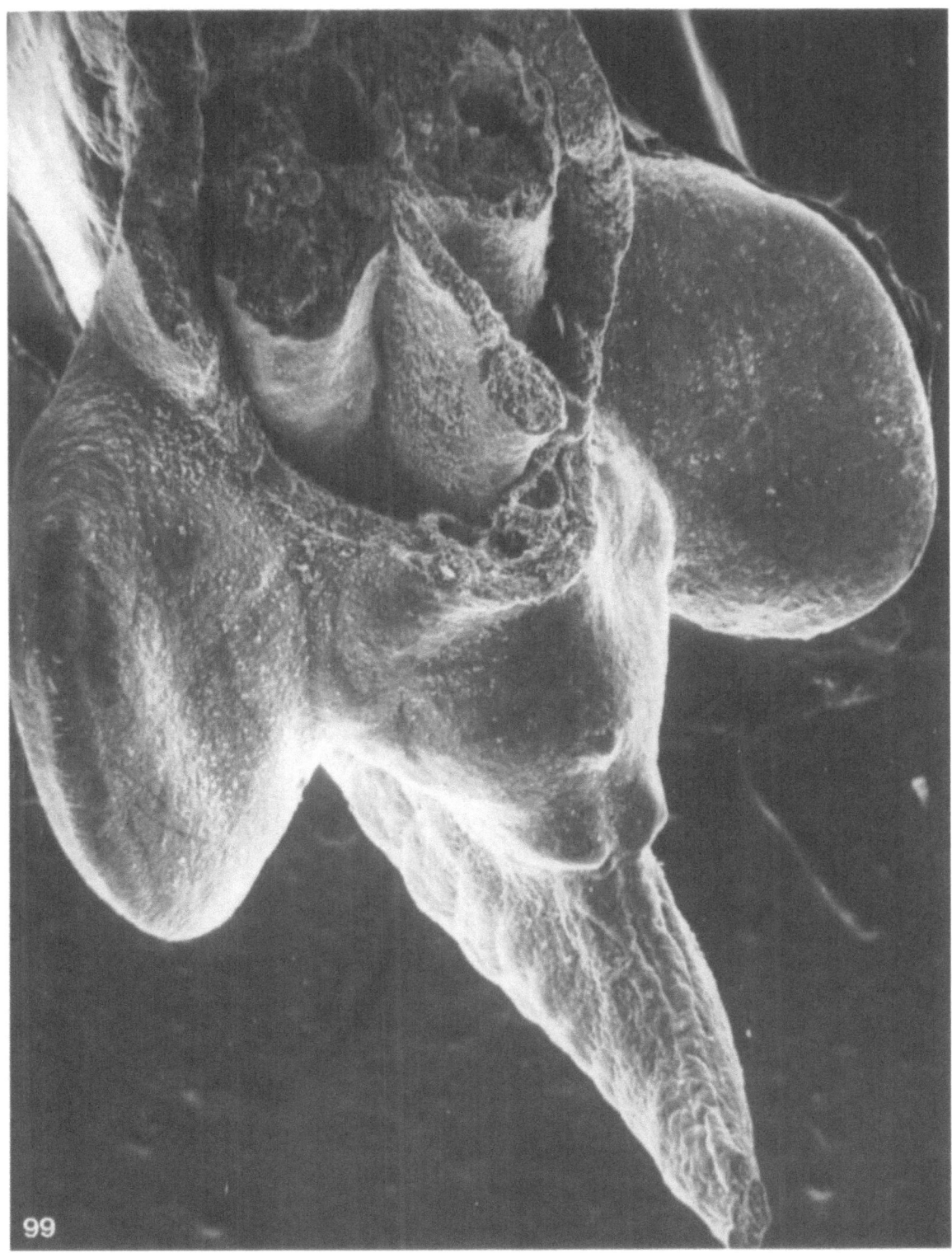

Figure 99. Transversely dissected lower portion of the body of an embryo. Stage 7–3. Structures depicted: urogenital ridges and mesentery with the gut protruding into the peritoneal cavity, body wall, urachus with laterally located umbilical arteries, and posterior limb buds. Indifferent external genitalia: anlage of glandular tubercle, anal hillocks, and tail of the embryo. (The depressed area underneath the glandular tubercle and between the anal hillocks is occupied by the urogenital membrane.)

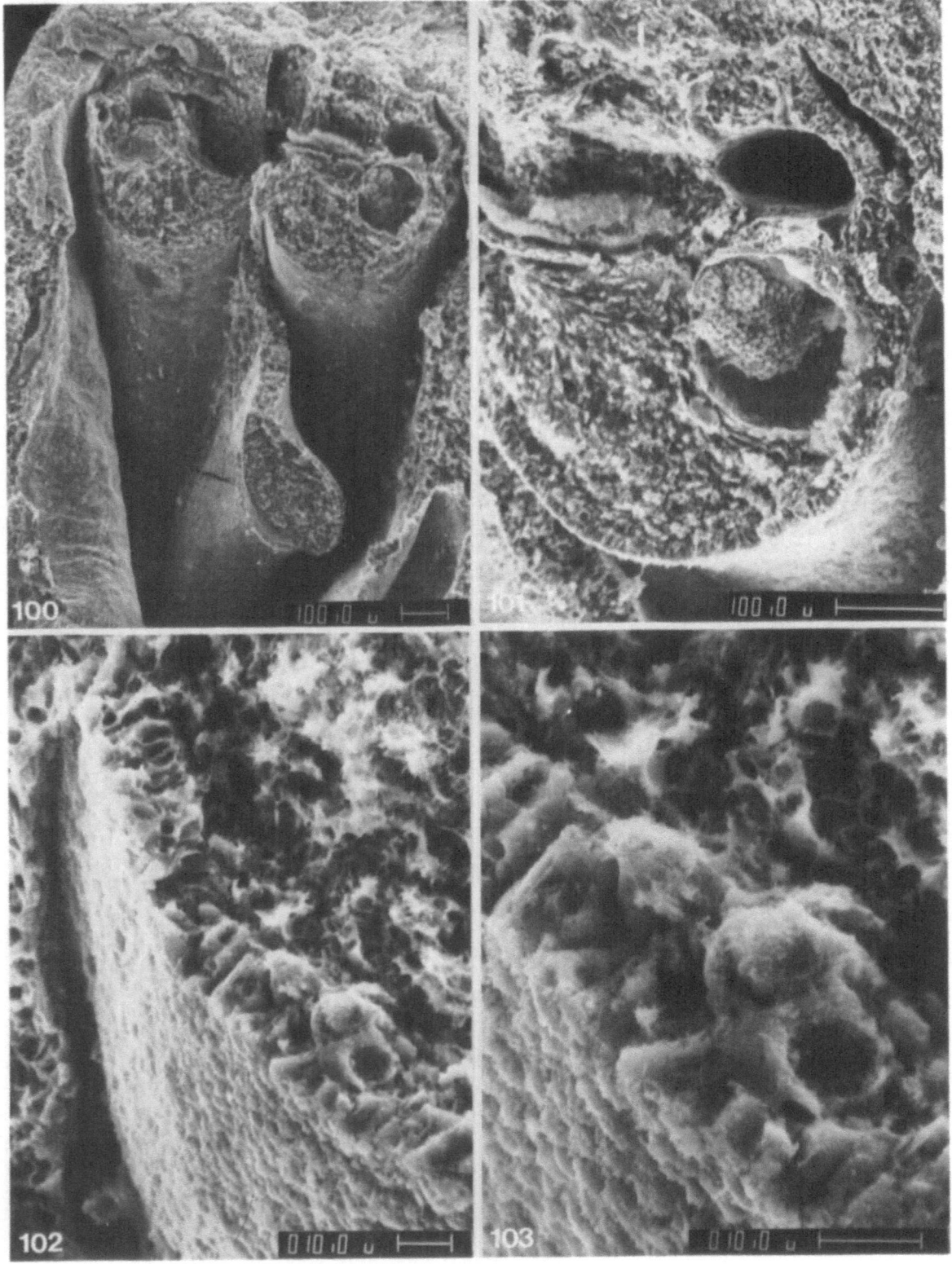

Figures 100–103. (100) Dissected urogenital ridges and mesentery with the gut. Stage 7-4. **(101)** Dissected urogenital ridge. The following structures are evident: surface mesodermal layer, mesenchymal stroma, mesonephric malpighian corpuscle, lumen of the posterior cardinal vein, and mesonephric duct. **(102)** Components of the dissected medioventral portion of the urogenital ridge: surface coelomic (mesodermal) epithelium, mesenchymal stroma, and primordial germ cells. **(103)** Dissected urogenital ridge. Big spheroid cells underneath the mesodermal epithelium are considered as primordial germ cells.

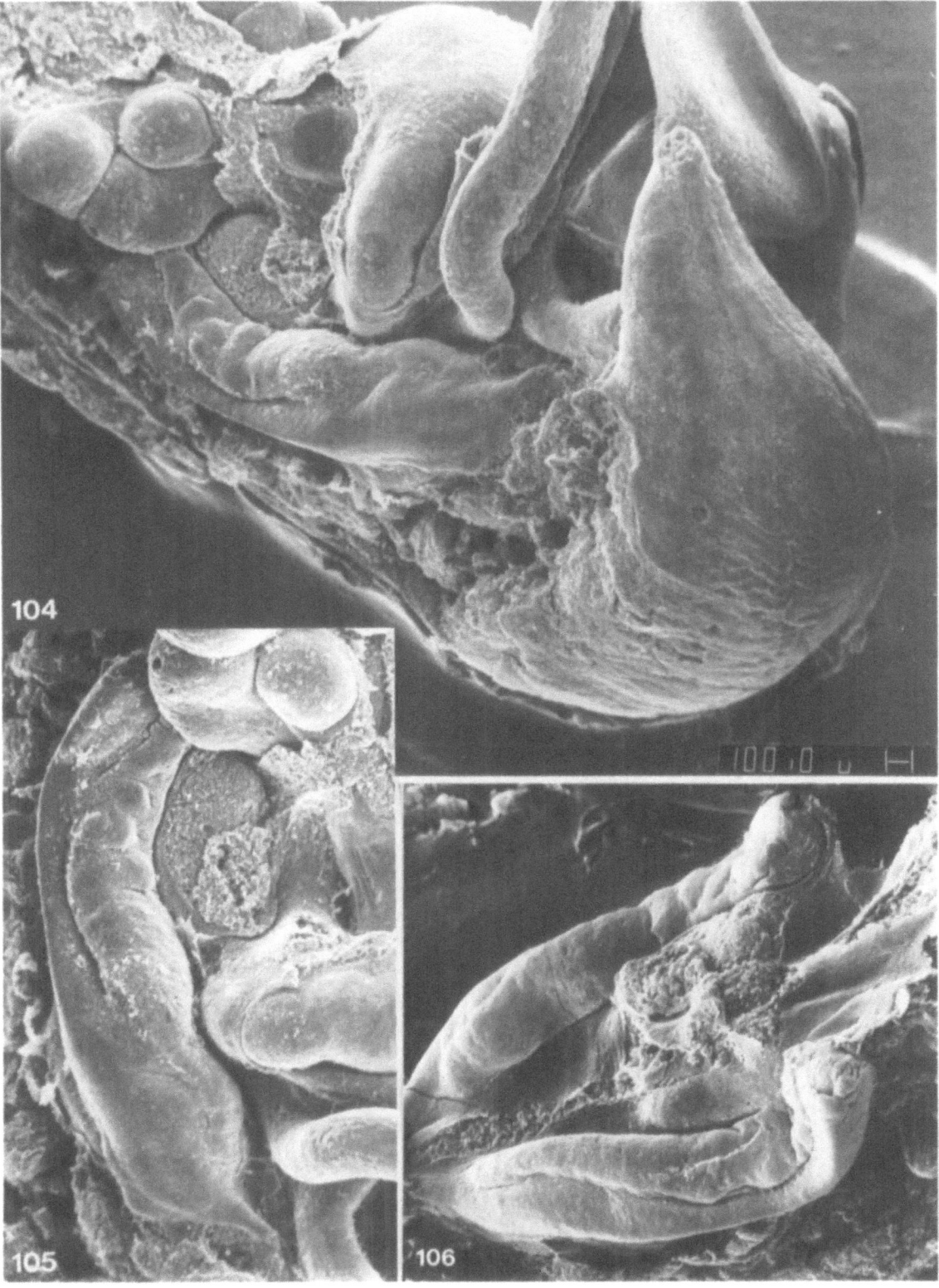

Figures 104–106. Differentiation of the urogenital ridges into mesonephric ridges and genital ridges. **(104)** Anatomy of visceral organs. Stage 7-4. Structures depicted: right lung with three lobes (left), dissected adrenal, stomach, gut, and tail of the embryo. The urogenital ridge with the developing genital ridge extends from the lung to the tail of the embryo. **(105)** The urogenital (mesonephric) ridge with a developing genital ridge. The genital ridge appears on the medioventral surface of the urogenital ridge. The urogenital ridge transforms into a genital and a mesonephric ridge. **(106)** Genital and mesonephric ridges dissected together with the dorsal body wall. Attachment of the removed mesentery is evident in the midline.

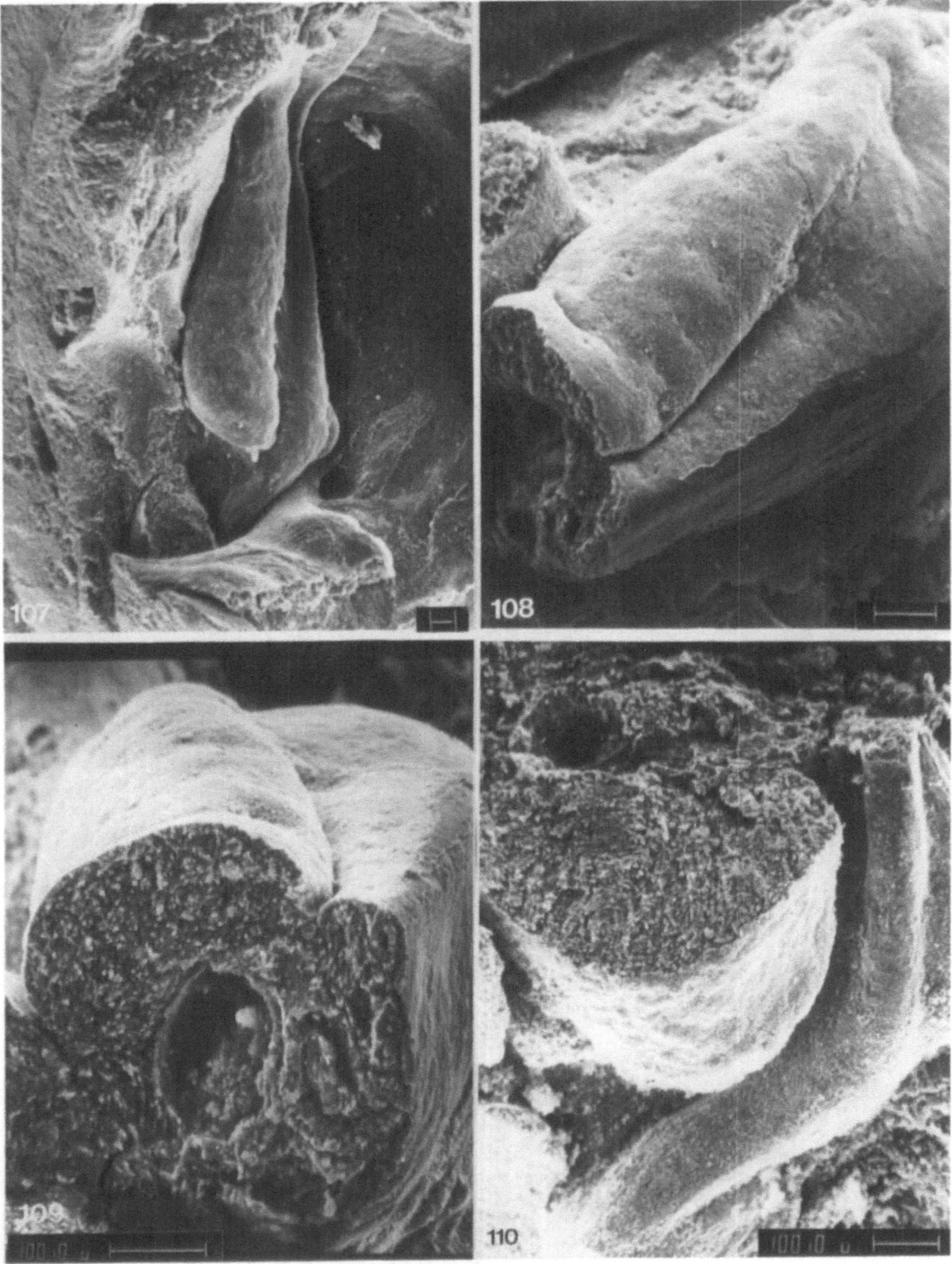

Figures 107–110. Formation of a gonad (probably a testis). Stage 7-5. **(107)** Gonad (genital ridge) on the medioventral surface of the mesonephric ridge. **(108)** Upper portion of the dissected genital and mesonephric ridges. **(109)** Dissected genital and mesonephric ridges. Close relationship of the gonadal blastema to the mesonephros is evident. **(110)** Lower portion of dissected gonad (probably a testis) and its relationship to the genital cord containing mesonephric and paramesonephric ducts.

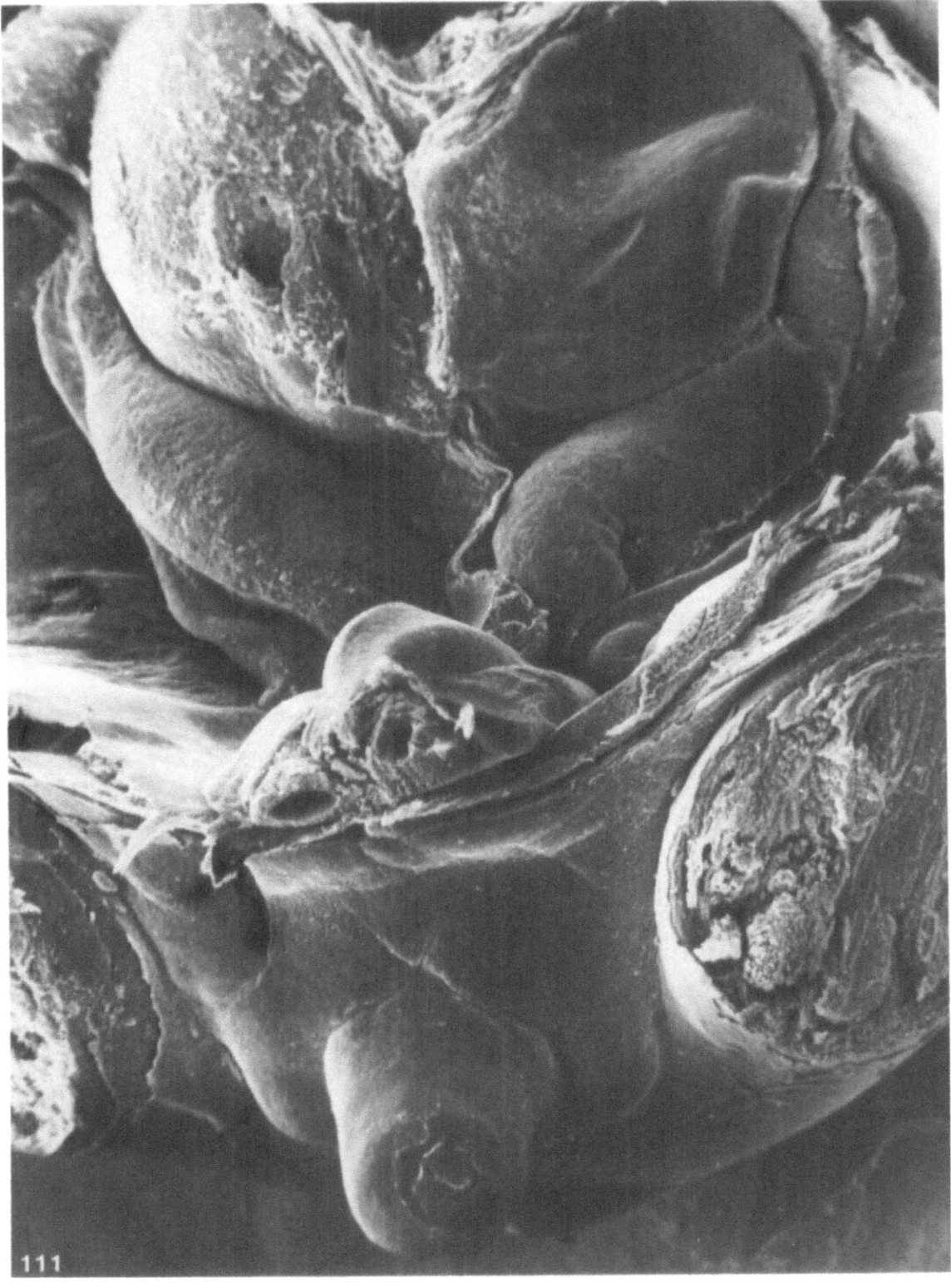

Figure 111. Anatomy of adrenals, gonads, regressing mesonephric ridges (containing mesonephric and paramesonephric ducts), urinary bladder, and external genitalia. Stage 8-1.

74

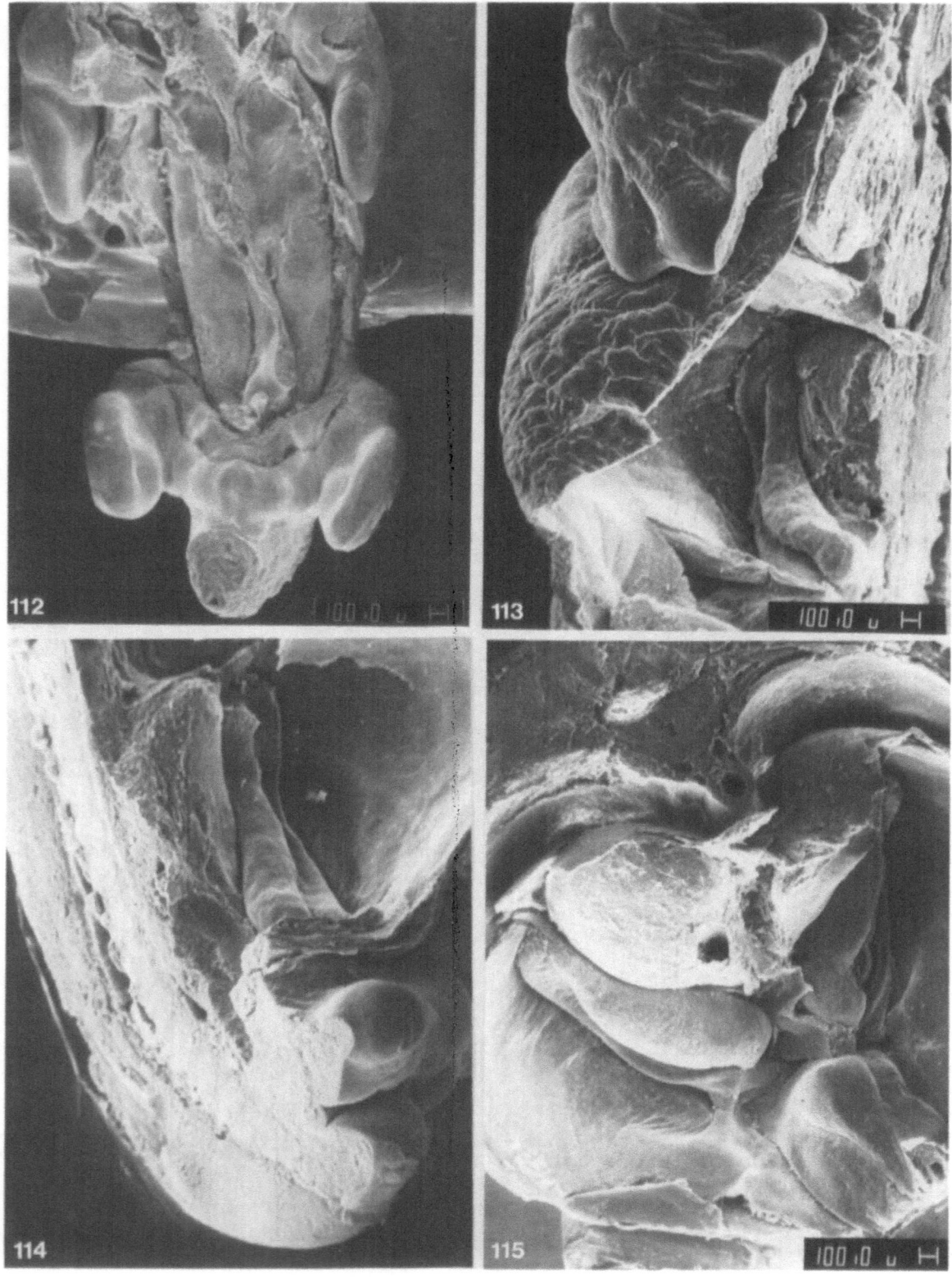

Figures 112–115. Early positional changes of gonadal primordia. **(112)** Longitudinally oriented urogenital ridges extending from the thoracic region to the sacral region. Stage 7-4. **(113)** Genital and mesonephric ridges extending from the diaphragm to the bottom of pelvis. Stage 7-5. Lateral displacement of both ridges is related to the development of adrenals and kidneys. **(114)** Genital ridge, mesonephric ridges, and external genitalia. Stage 7-5. **(115)** Anatomy of abdominal organs. Stage 8-1. Gonads, mesonephric ridges, adrenals, mesentery with the gut, and urinary bladder. Note the relative descensus of gonads, ascensus of adrenals, and positional changes related to the development of the adrenals.

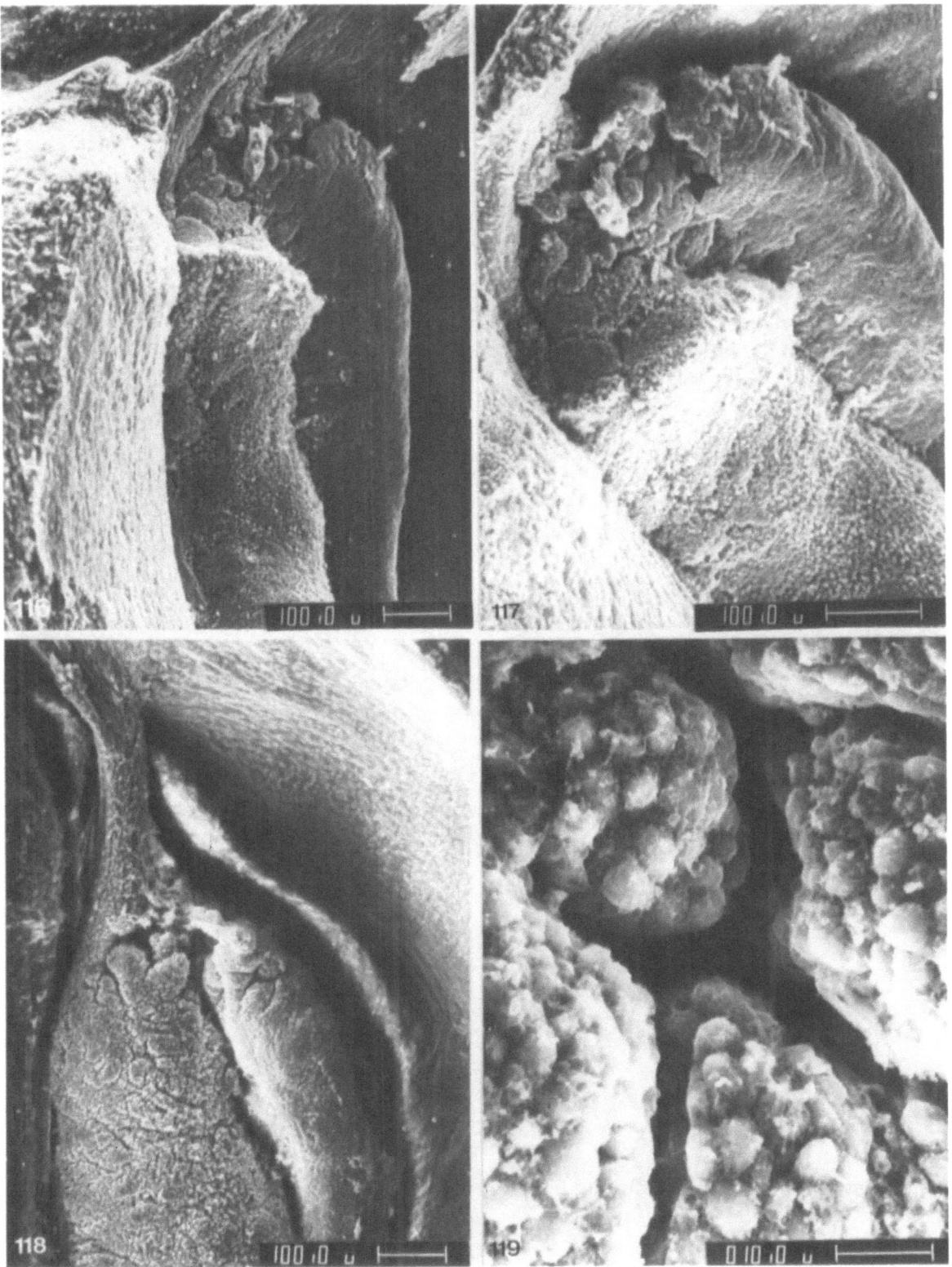

Figures 116–119. Development of the paramesonephric (Müllerian) duct. **(116)** Cranial ends of mesonephric and gonadal ridges. Stage 7-5. **(117)** 'Müllerian area characterized by irregular papillarities involves cranial areas of both ridges. Stage 7-5. The paramesonephric (Müllerian) epithelium invaginates (arrow) laterodorsally along the cranial obliterated end of mesonephric (Wolffian) duct. **(118)** Cranial portion of the genital ridge (the embryonic ovary) and mesonephric ridge with the opening of paramesonephric duct. Stage 8-1. Attachment of both ridges to the body wall by the cranial ligament of the urogenital ridge (the future proper ovarian ligament). **(119)** Abdominal opening of paramesonephric duct (the future infundibulum of the oviduct; detail from the previous figure).

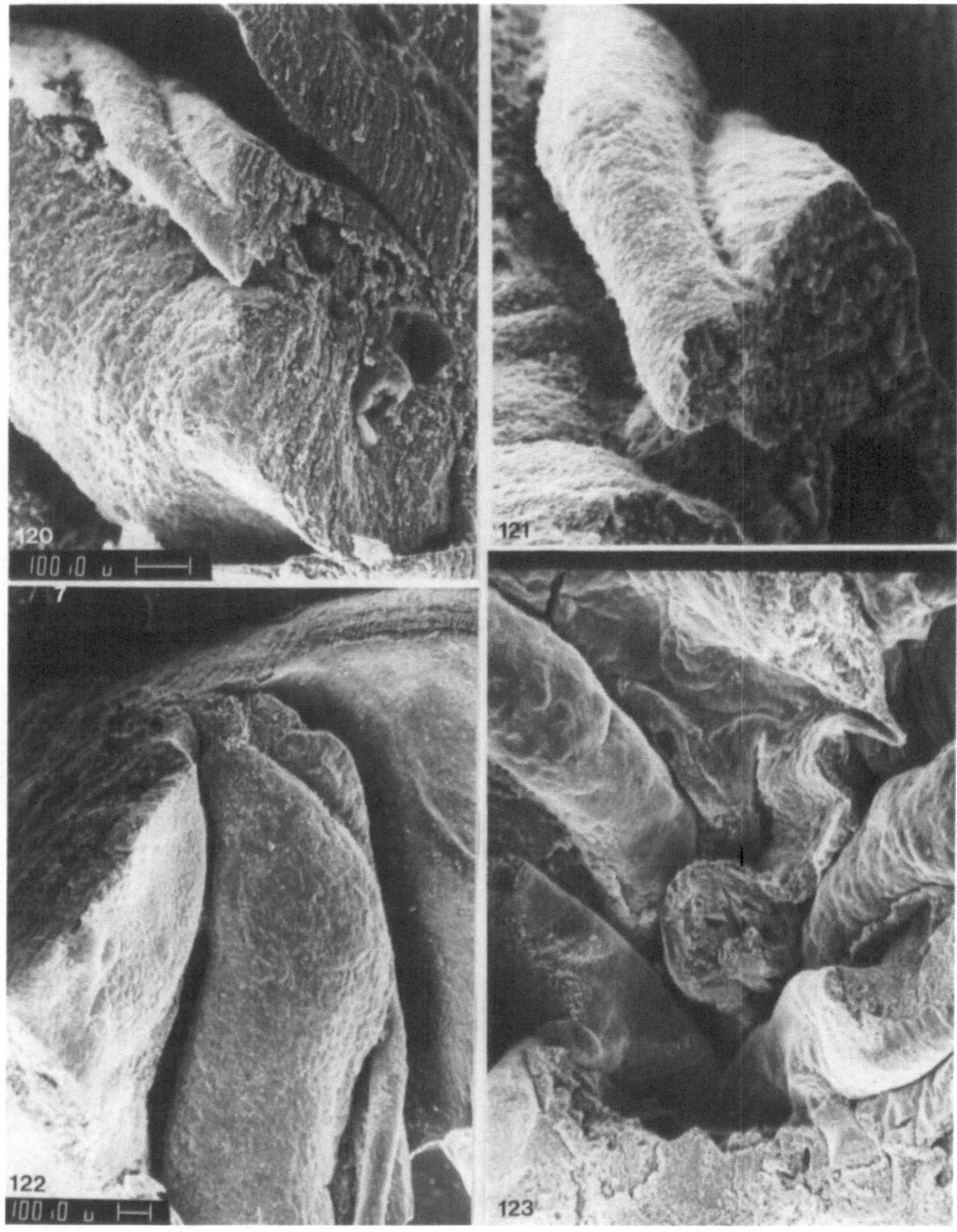

Figures 120–123. (120) Crossing of paramesonephric (Müllerian) and mesonephric (Wolffian) ducts. Stage 7-5. The lower portion of paramesonephric duct becomes located medially to the mesonephric duct. **(121)** Crossing of paramesonephric (medially) and mesonephric (laterally) ducts. **(122)** Cranial end of paramesonephric duct located close to the upper pole of gonad. The duct closes in male embryos, but remains open in females. **(123)** Dissected pelvic organs. Stage 8-2. The gut attached by the mesentery to the body wall is interposed between ovaries. Fusing genital cords containing paramesonephric and mesonephric ducts are attached to the inguinal areas by caudal mesonephric ligaments (the future testicular gubernaculum in males, the round ligament of uterus in females).

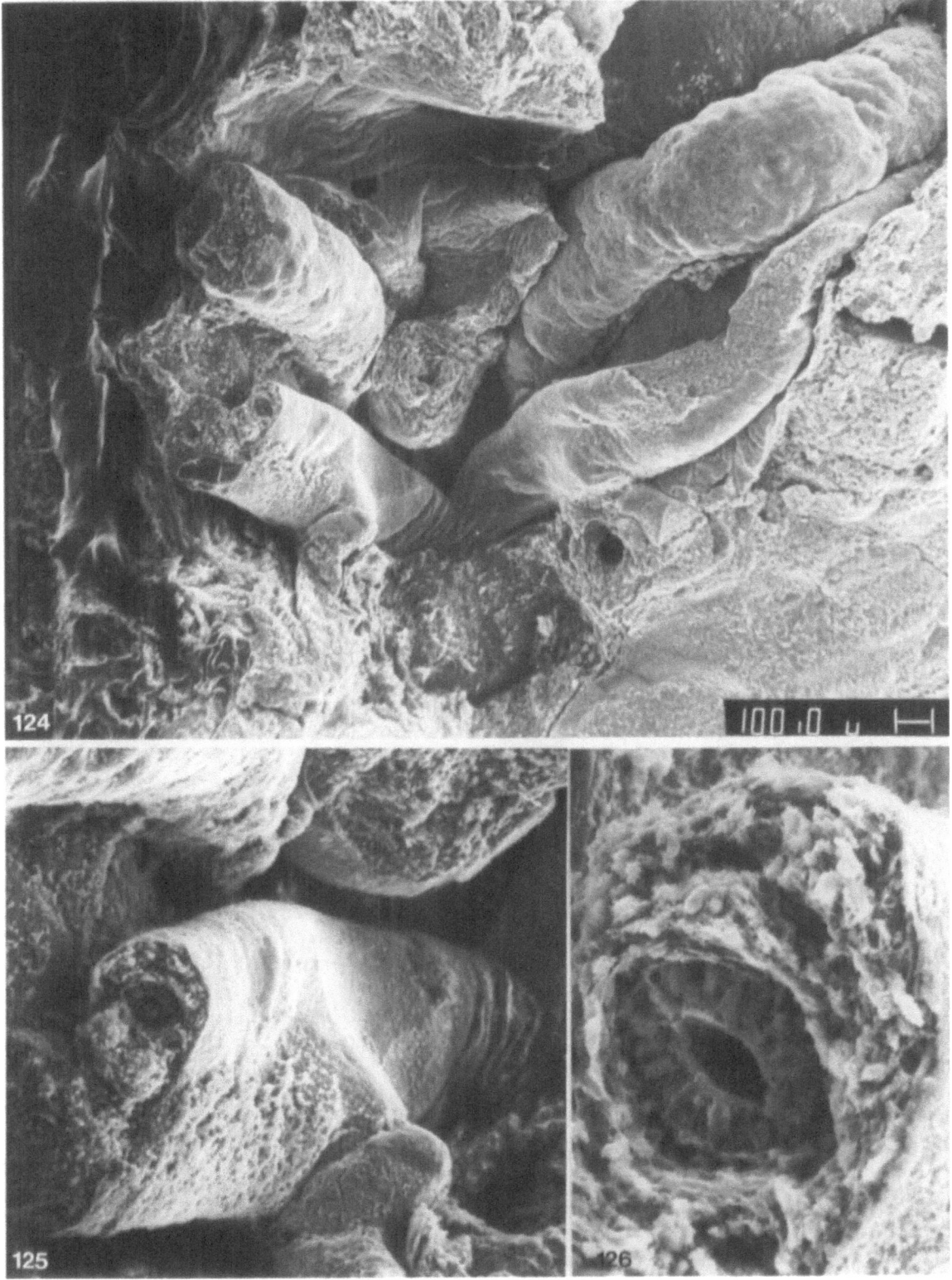

Figures 124–126. (124) Dissected pelvic organs. Stage 8-2. Gut with the mesentery interposed between ovaries. Fusing genital cords containing genital ducts are attached to the inguinal areas by round uterine ligaments. Position of the paramesonephric (located medially) and mesonephric (located laterally) ducts within the genital cord is evident on the dissected (right) side. **(125)** Detail of the paramesonephric duct and its attachment by the round ligament. **(126)** Paramesonephric duct lined by a single-layered cylindrical epithelium.

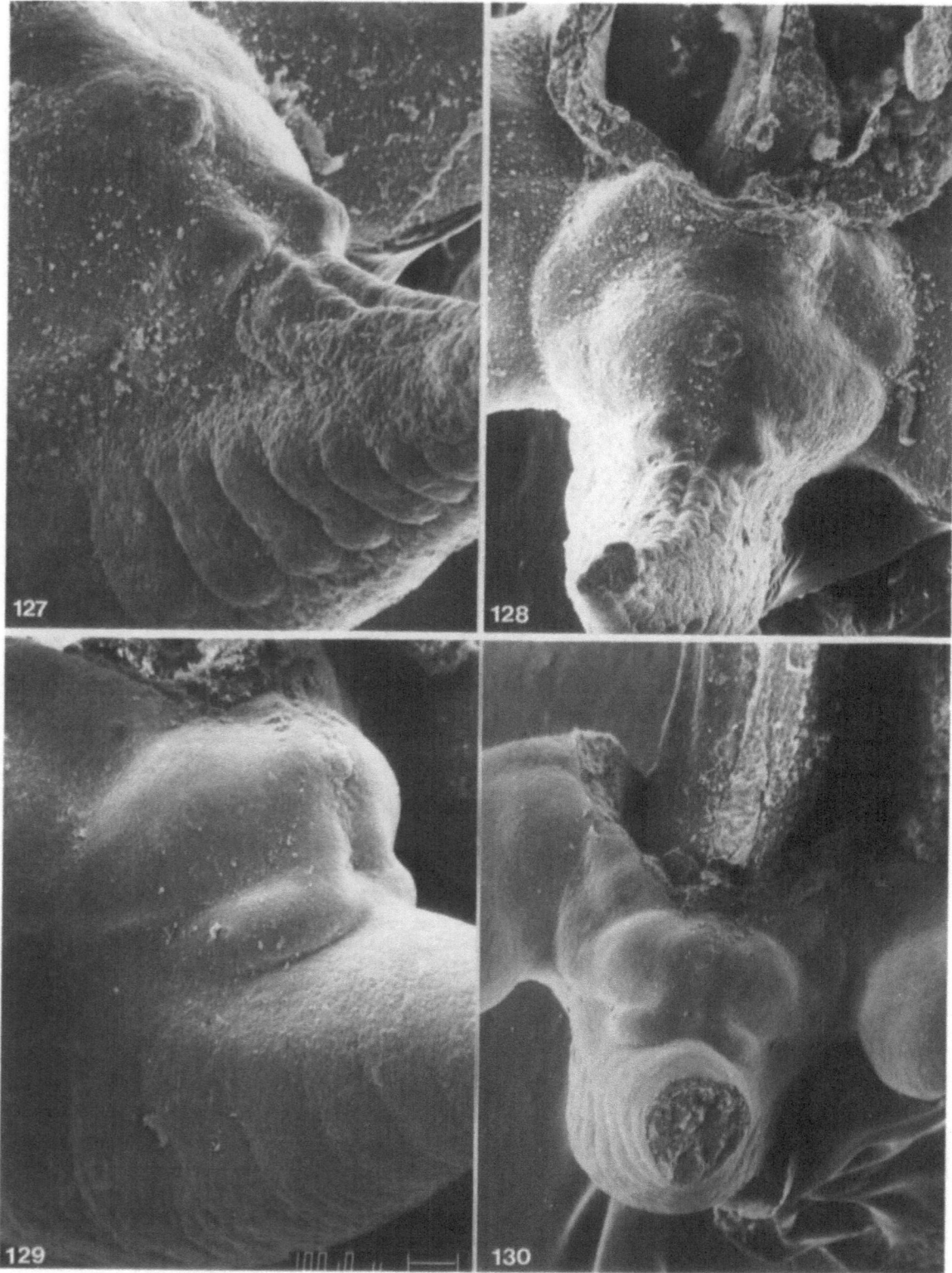

Figures 127–130 (127) Anlage of external genitalia. Stage 7-3. Structures depicted: glandular tubercle and paired anal hillocks and tail folds. The depressed area underneath the genital tubercle, between the anal hillocks and above the tail folds, is occupied by the cloacal membrane. Oblique lateral view. **(128)** Frontal view of an anlage of the external genitalia. Stage 7-3. The area is delineated by the attachment of the umbilical cord, posterior limb buds, and the tail. **(129)** Indifferent external genitalia. Early stage 7-4. Sagitally oriented, depressed cloacal membrane delineated by urethral folds with the glandular tubercle anterior and anal hillocks posterior is located on the ventral side of genital tubercle. Oblique lateral view. **(130)** Frontal view of the same specimen as in Figure 129. Genital tubercle of the classic embryology is composed of a glandulourethral primordium (enclosing the primitive glandular tubercle) and two corporal primordia located superolaterally.

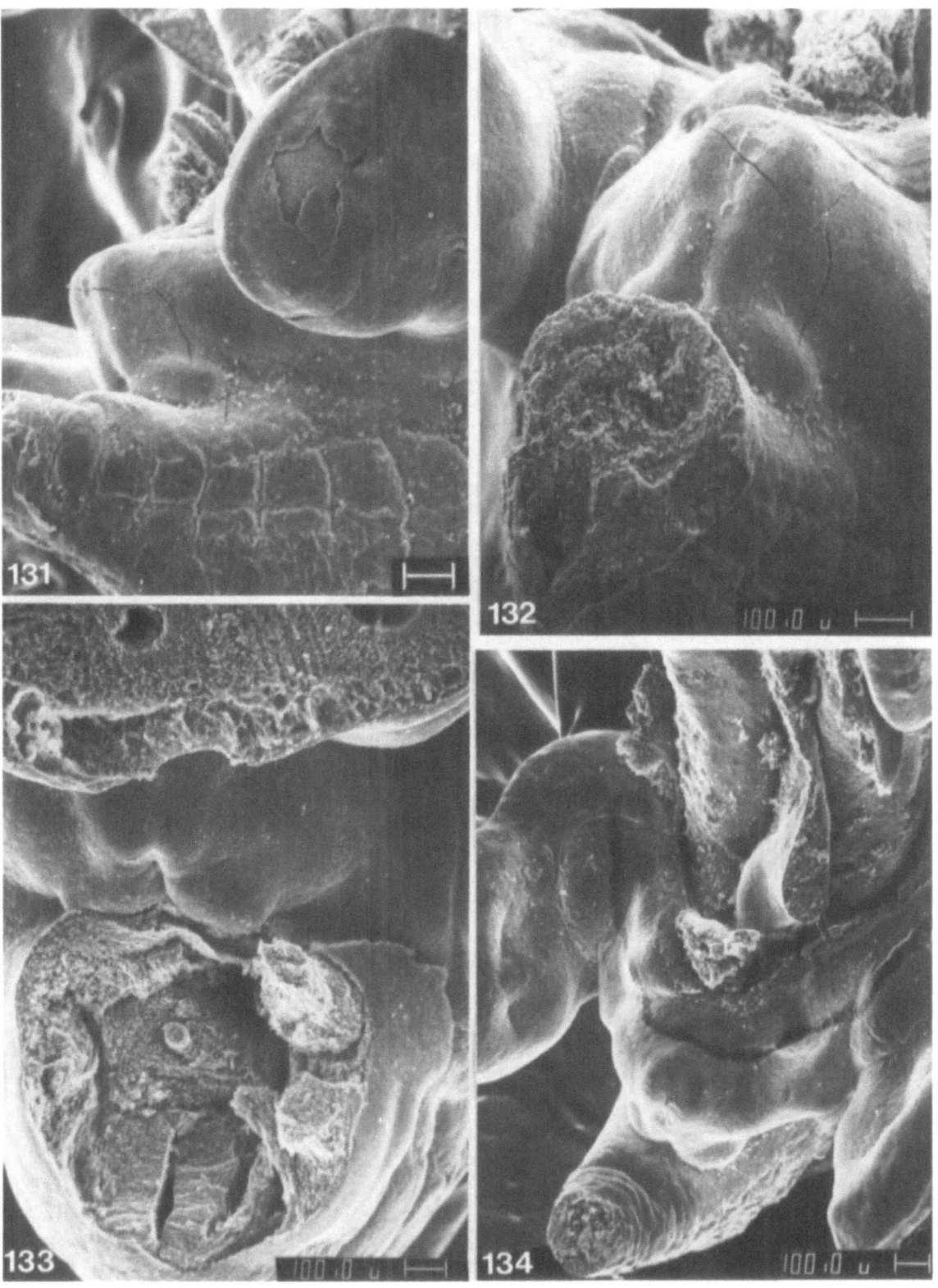

Figures 131–134. (131) External genitalia. Stage 7-4. Genital tubercle with sagitally oriented cloacal membrane and anal hillocks. Lateral view. **(132)** External genitalia. Stage 7-4. Genital tubercle with sagittally oriented cloacal membrane delineated by the urethral plate and anal hillocks. Glandular and corporal portions of the cloacal membrane, and the urethral plate, are evident. **(133)** Indifferent external genitalia. Stage 7-5, located between the cut umbilical cord and the embryonic tail (removed). **(134)** Indifferent external genitalia. Stage 7-5.

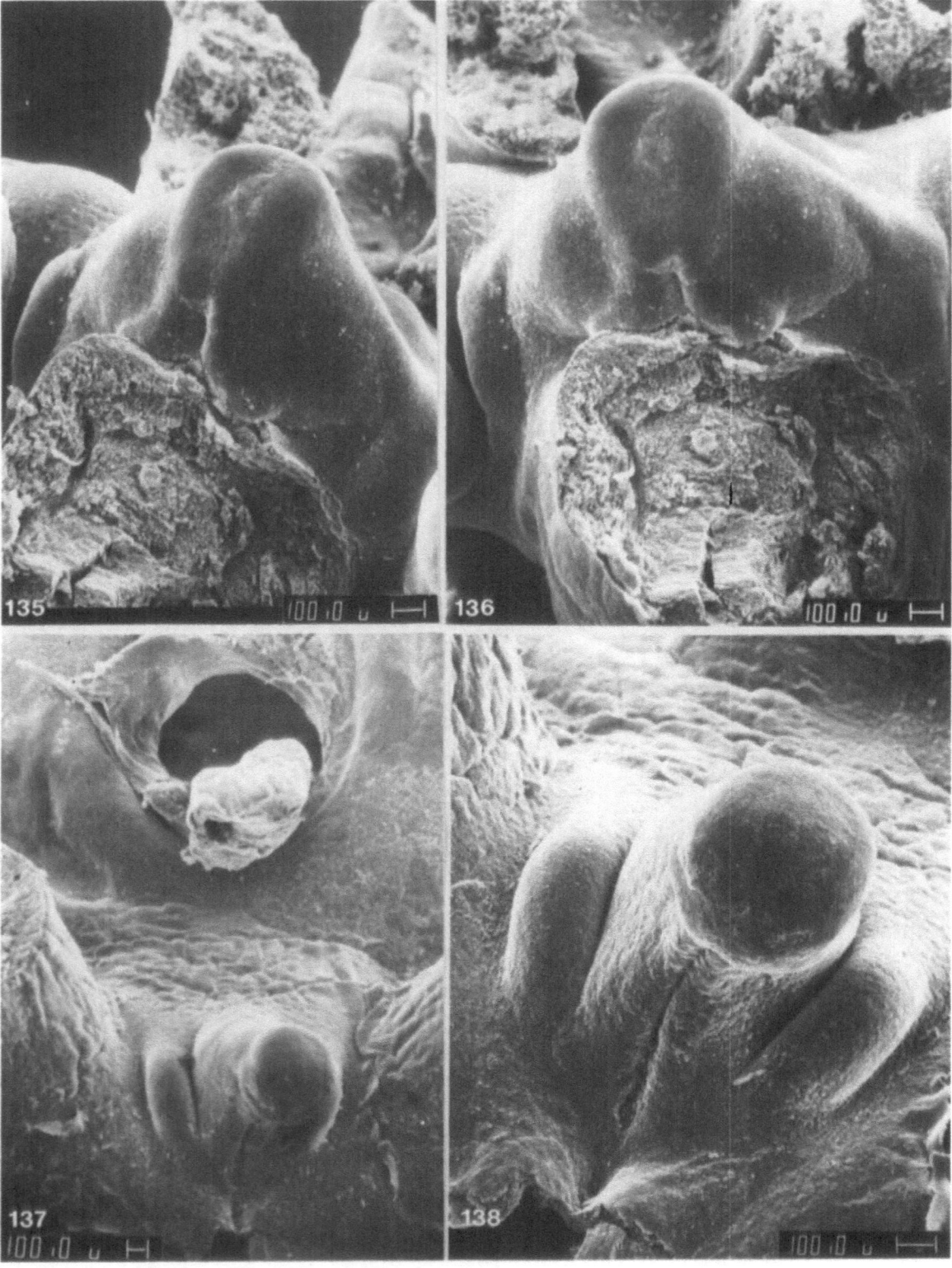

Figures 135–138. (135) Indifferent external genitalia. Stage 7-5. Genital tubercle consists of an unpaired urethroglandular portion and a urethrocorporal portion (urethral plate) located around the cloacal membrane, and paired cavernous corporal portions. Primordia of labioscrotal swellings are lateral to the genital tubercle. Oblique view. **(136)** Anterior view of the same specimen as on Figure 135. Epithelial plug on the tip of the glandular primordium represents the most anterior portion of the cloacal membrane. **(137)** Anterior body wall. Stage 8-1. Umbilical cord removed and urachus present within the opening into the umbilical coelom. Indifferent external genitalia. **(138)** Indifferent external genitalia. Stage 8-1. Derivative of the genital tubercle, the phallus exhibits a glandular and a corporal portion (with a deep urethral groove). Primitive perineum (urorectal septum) separates the urogenital membrane and the anal membrane. Labioscrotal swellings are distinct.

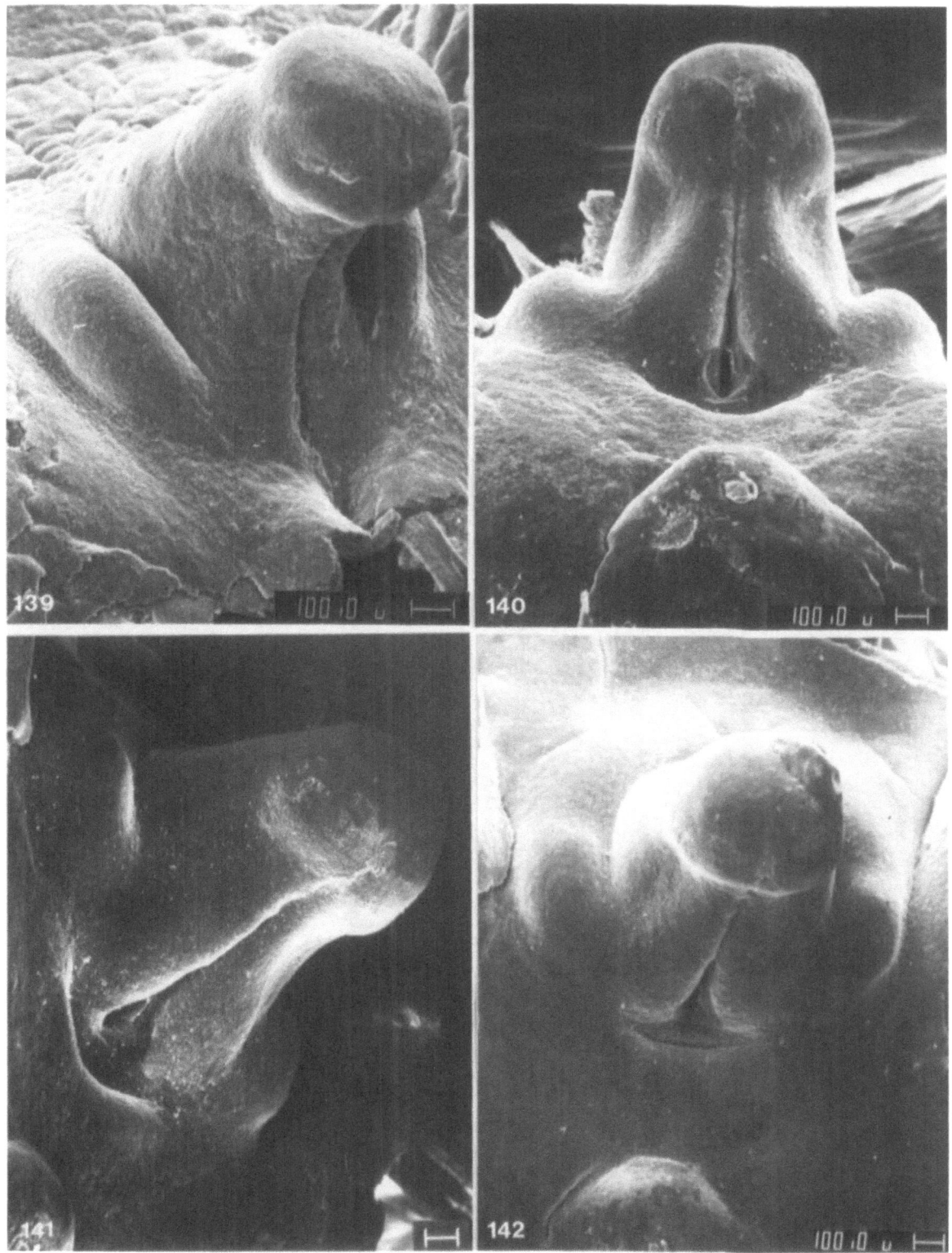

Figures 139–142. (139) Indifferent external genitalia. Stage 8-1. The phallus exhibits an glandular portion and a corporal portion. The urethral groove is located on the inferior side of the corporal portion. Labioscrotal swellings appear laterally from the phallus. **(140)** Indifferent external genitalia. Stage 8-2. Inferior frontal view. Structures depicted: phallus with a glandular primordium and a corporal primordium, urethral groove with external opening of the urogenital sinus, and labioscrotal swellings. **(141)** Indifferent external genitalia. Stage 8-2. Inferior oblique view. **(142)** Indifferent external genitalia. Stage 8-2. Anterior frontal view. Note the primitive perineum separating the external opening of the urogenital sinus from the anal opening.

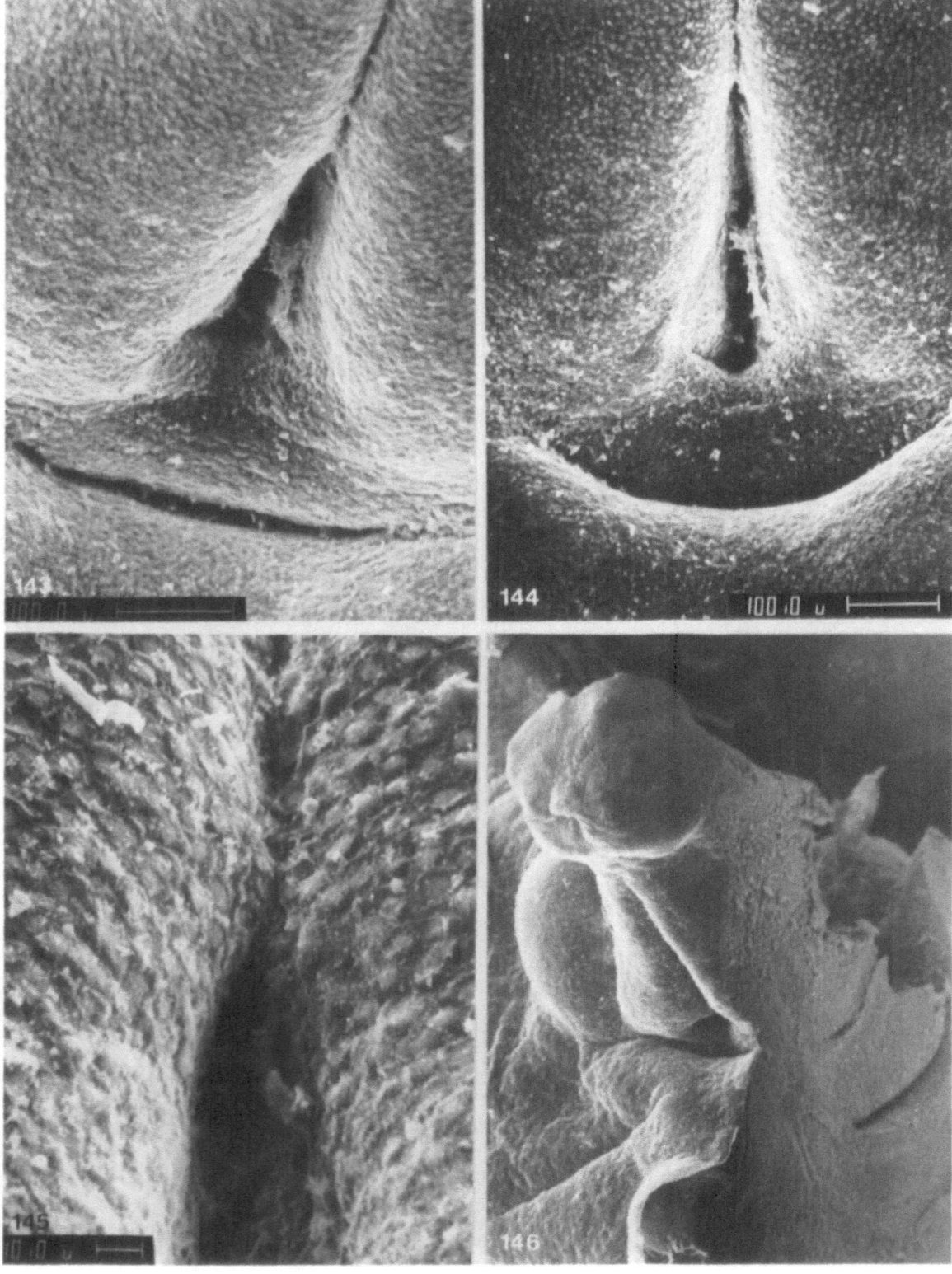

Figures 143–146. (143 and 144) External opening of the urogenital sinus, primitive perineum, and anal opening. Stage 8-1. **(145)** Surface of the closing urethral groove. **(146)** Dissected external genitalia at the end of the indifferent period. Stage 8-2. Structures depicted: phallus – glandular portion with an epithelial plug, corporal primordium; right labioscrotal swelling, primitive perineum, and anus.

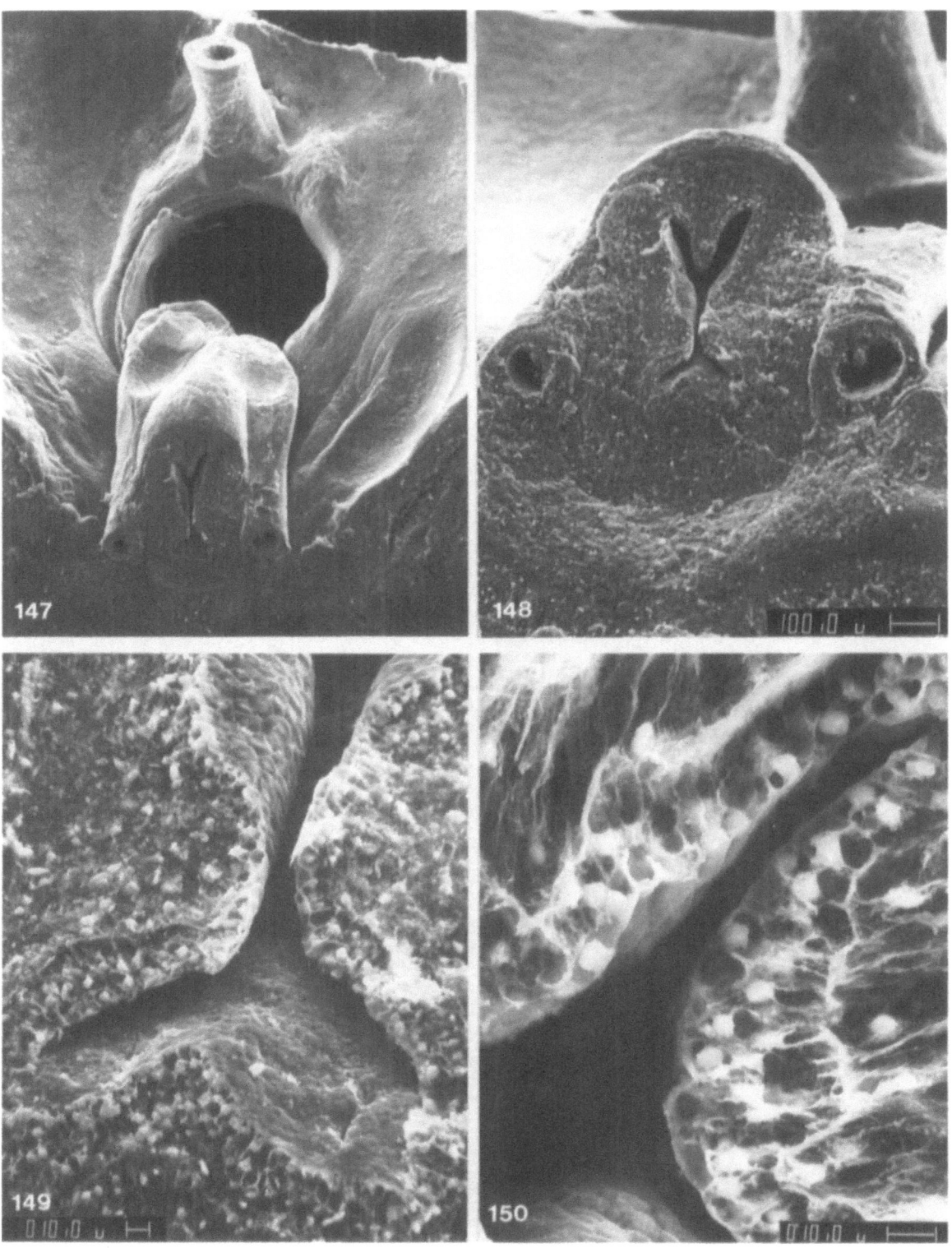

Figures 147–150. Urinary bladder. Stage 8-1. **(147)** Dissected anterior body wall from inside. Structures depicted: umbilical vein, communication of the peritoneal cavity and the umbilical coelom, and vertex of the dissected urinary bladder with two umbilical arteries (laterally). **(148)** Dissected urinary bladder and umbilical arteries. **(149)** Dissected wall of the urinary bladder. **(150)** Epithelium of the urinary bladder.

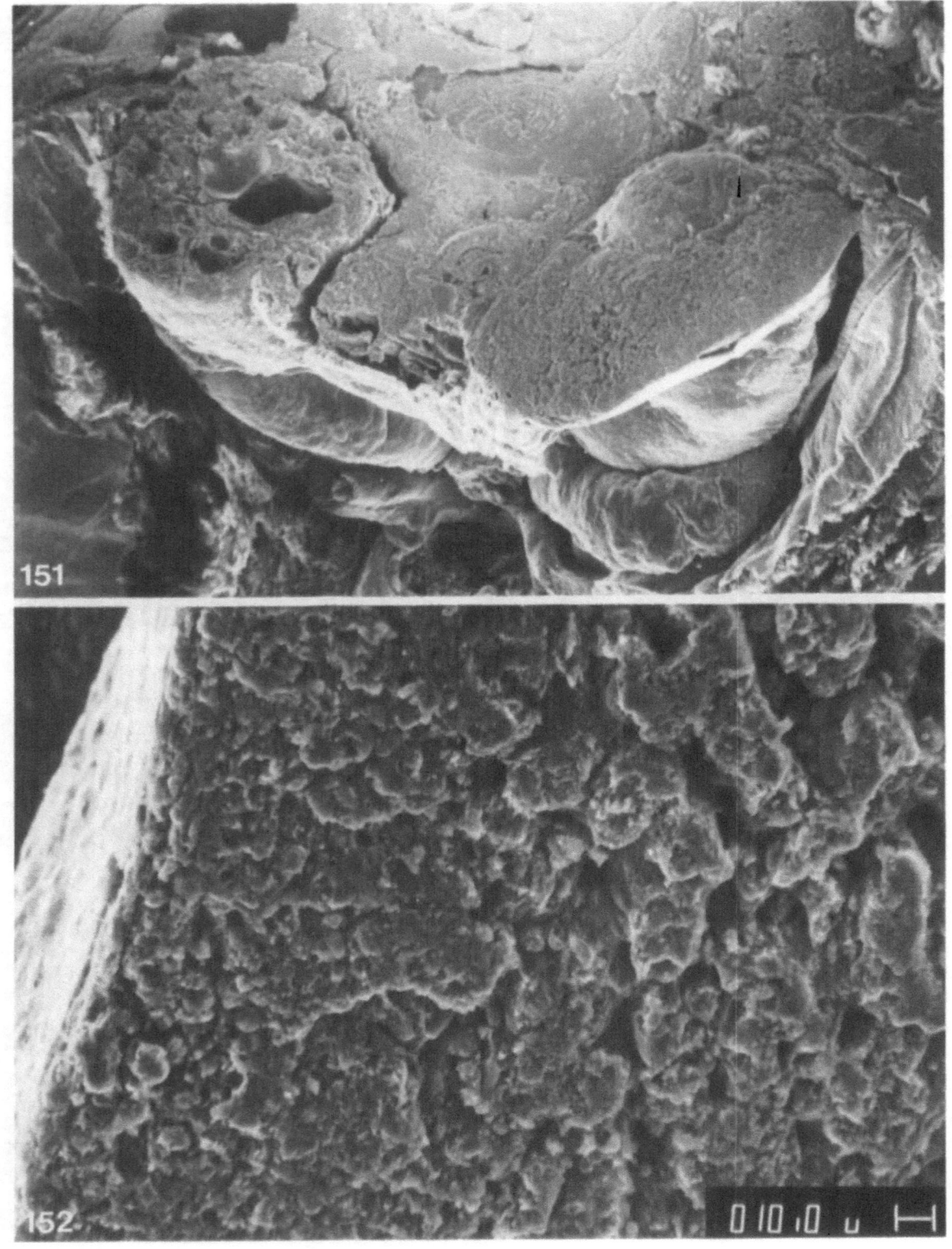

Figures 151 and 152. (151) Transverse dissection of fetal body. Stage 8-2. Dissected kidney on the right side and adrenal on the left side. **(152)** Dissected fetal adrenal. Stage 8-2.

5. ORGANS WITH ENDODERMAL COMPONENTS

The digestive organs

The digestive tube is derived from the dorsal (intraembryonic) portion of the yolk sac. In relation to the lateral delineation of the embryo and to the formation of the anterior body wall and the umbilical cord, the dorsal portion of the yolk sac transforms into the primitive gut, which is closed anteriorly by the buccopharyngeal membrane and posteriorly by the cloacal membrane. The blind anterior portion of the primitive gut is known as the foregut, the middle portion as the midgut, and the posterior portion as the hindgut. The allantois extends from the ventral wall of the hindgut into the connecting stalk. The midgut is connected by the vitelline (omphalomesenteric) duct with the yolk sac. The gut protrudes into (is located within) the coelomic cavity and is suspended to the dorsal body wall by the dorsal mesentery. In addition, the anterior gut is attached to the ventral body wall by the ventral mesentery. The respiratory organs and the liver evaginate from the foregut into the ventral mesentery. The pancreas develops from two evaginations: the ventral one is formed close to the liver and grows into the ventral mesentery; the dorsal one proliferates from the dorsal duodenal wall into the dorsal mesentery.

The intraabdominal portion of the digestive tube is supplied by three arteries: the coeliac artery brings blood to the derivatives of the anterior gut, the superior mesenteric artery supplies derivatives of the midgut, and the inferior mesenteric artery is destined for derivatives of the hindgut. The development of the most anterior portion of the foregut contributes to the oral cavity and transforms into the pharynx. The posterior portion of the foregut gives rise to the esophagus, to the stomach, and to the upper portion of the duodenum. The midgut contributes to the jejunoileum, the caecum with the appendix, and the transverse colon. Derivatives of the hindgut are the descending colon, the sigmoid, the rectum, the vesicourethral primordium, and the urogenital sinus.

The pharynx and brachiogenic pharyngeal complex (thyroid, parathyroids, palatinal tonsils, and thymus) (Figs. 153 and 154).
The early embryonic pharynx is a funnelshaped ventrodorsally flattened space lined by endoderm and located behind the pharyngeal membrane. The rostral portion of the notochord is incorporated in the midline within the pharyngeal ceiling. A small invaginated area where the notochord leaves the endoderm and enters the mesenchyme shortly before its attachment to the posterior wall of the Rathke pouch is known as the Seesel diverticulum. The ventrolateral pharyngeal walls are segmented into the pharyngeal arches and pouches externally covered by the ectoderm and internally covered by the endoderm. The pharyngeal floor is formed by fusing pharyngeal arches, by a midline mesenchymal elevation known as the tuberculum impar, located between the first and second pharyngeal arches, and by the copula (or the hypobranchial eminence) located between the third, fourth, and fifth pharyngeal arches. Each pharyngeal arch is supported by cartilage and contains a proper nerve and an aortic arch. Phylogenetically, the pharyngeal (branchial) arches were formed as the skeleton supporting the gills, a respiratory organ in fish. With the phylogenic transition of animals from water to earth, the endoderm of the pharyngeal pouches was transformed into lymphatic organs and some endocrine glands. The floor of the pharynx contributed to the tongue.

Derivatives of the endodermal first pharyngeal pouch are the middle-ear cavity and the nasopharyngeal tube. The ectodermal portion of the

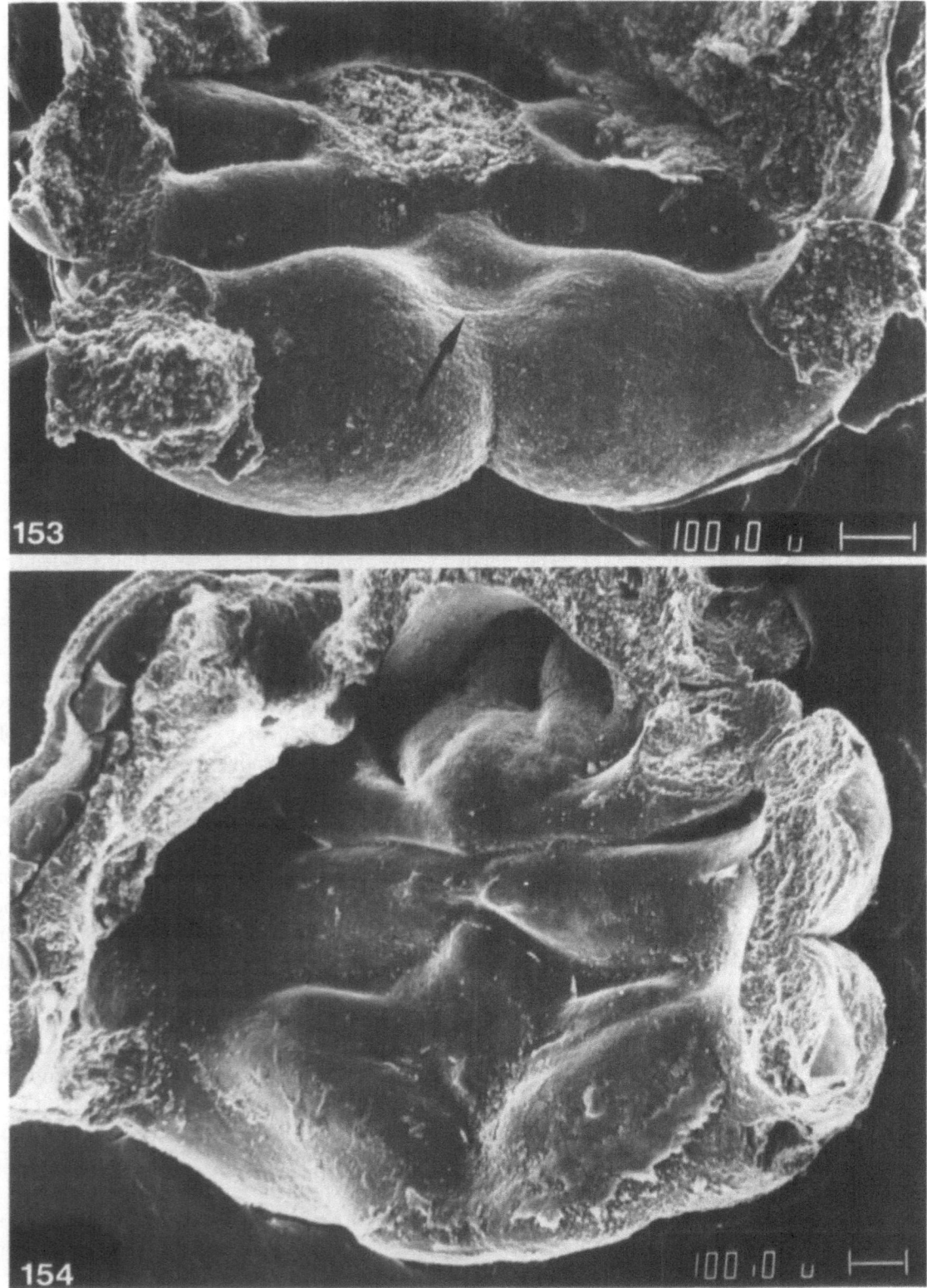

Figures 153 and 154. Dissections of pharyngeal floor. **(153)** Floor of the primitive oral cavity and pharynx. Stage 6-3. Structures depicted: mandibular arch composed of two fusing first pharyngeal arches. (A discrete transverse sulcus evident on the surface – arrow – is regarded as a remnant of the insertion of the regressed pharyngeal membrane); impar tubercle (unifying posteriorly first pharyngeal arches); first, second, and third pharyngeal pouches (ventral portions); and second, third, and fourth pharyngeal arches fusing into a copula (surface of copula artificially damaged). **(154)** Dissected floor of the pharynx. Stage 7-2. Structures depicted: mandibular arch with two lateral lingual swellings and tuberculum impar, second (hyoid) pharyngeal arches detached from the copula; the copula unifying pharyngeal arches 3, 4 and 5; and pharyngeal pouches 1, 2, 3, and 4.

first pharyngeal pouch contributes to the external auditory meatus. The bilaminar ecto- endodermal obturatory membrane of the first pharyngeal pouch contributes to the eardrum.

The endodermal second pharyngeal pouch provides the tonsillar fossa for the palatial tonsil, and the tonsillar cryptae. The third and fourth endodermal pharyngeal pouches each differentiate into a dorsal portion and a ventral portion. The ventral portion of the third pharyngeal pouch represents the primordium of the thymic reticulum. The dorsal portion of the third pouch gives rise to the inferior parathyroid (descending with the thymus). The dorsal portion of the fourth pouch changes into the superior parathyroid. The ventral portion of the fourth pharyngeal pouch, considered by some as the fifth pharyngeal pouch, contributes to the lateral primordium of the thyroid gland.

The tongue (Figs. 76c–78c and 153–158). The tongue develops from mesenchymal condensations of the floor of the primitive pharynx. The body of the tongue is formed from a midline eminence, known as the tuberculum impar, located between and behind the first pharyngeal arches, which fuses with two lateral lingual swellings derived from the posterior portions of mandibular arches. The lateral lingual swellings are located underneath the insertion of the pharyngeal membrane. The root of the tongue is derived from the second pharyngeal arches, originally incorporated in the copula. After detachment of the second arches, the copula is a midline mesenchymal elevation joining the third, fourth, and fifth pharyngeal arches. The copula contributes to the epiglottal swelling. The musculature of the tongue originates from occipital myotomes. The border between the corpus of the tongue (derived from the lateral lingual swellings of the mandibular arches and the tuberculum impar) and the radix (a derivative of the second arches) is evident as the sulcus terminalis. An endodermal area posterior to the tuberculum impar invaginates as the medial primordium of the thyroid gland. This site can be recognized in some individuals as the foramen caecum of the tongue.

The thyroid. The thyroid gland originates from three endodermal primordia: one main, or medial, and two accessory, or lateral. The main primordium invaginates from the endoderm of the floor of the primitive oral cavity in the midline between the first and second pharyngeal arches, and between the future impar tubercle of the tongue and the copula, respectively.

The medial thyroid invagination becomes bilobed and detaches from the tongue (the foramen caecum). The lateral (accessory) thyroid primordia are derived from the endodermal fourth pharyngeal pouches. In absence of the medial thyroid, the lateral primordia fail to develop. About three-fourths of the definitive gland develops from the main primordium.

The esophagus
The embryonic esophagus is a narrow tube connecting the pharynx with the stomach.

The stomach (Figs. 159–161)
The primitive stomach is a fusiform, longitudinally oriented dilatation of the anterior gut. The greater curvature attached by the dorsal mesentery is originally oriented dorsally. The lesser curvature is oriented ventrally and is attached to the septum transversum (the primordium of the diaphragm) by the ventral mesentery. The primitive stomach rotates around the longitudinal axis and the antero-posterior (sagittal) axis. By the 90° relation along the longitudinal axis, the greater curvature faces the left side and the lesser curvature faces the right side. The left surface becomes the anterior surface and the right becomes the posterior surface (the left nervus vagus goes to the anterior surface). The rotation around the sagittal axis brings the pyloric end up and to the right and the cardiac portion to the left. The dorsal mesogastrium transforms into the greater omentum and the ventral mesogastrium, between the liver and the stomach, becomes the lesser omentum.

The duodenum (Fig. 104)
The duodenum is caudal to the stomach. The hepatic diverticulum grows from the terminal portion of the foregut into the ventral mesentery (mesoduodenum). The convexity of the duodenal loop is originally oriented anteriorly and later rotates to the right.

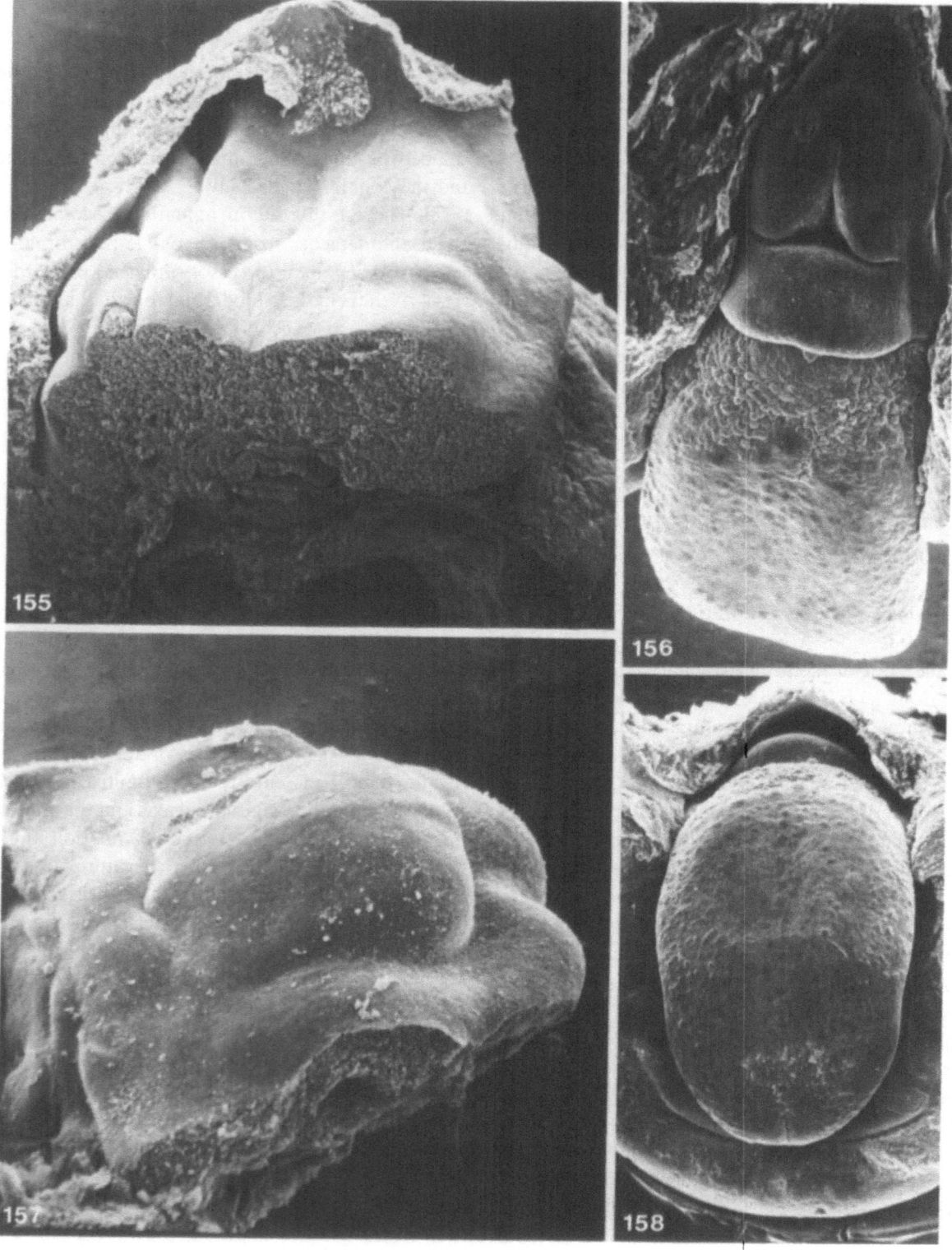

Figures 155–158. The tongue. **(155)** Parasagittally dissected floor of the oral cavity and pharynx. Stage 7-4. Structures depicted: mandible with lower lip and body of tongue comprising lateral lingual swellings and impar tubercle; radix of tongue comprising the second pharyngeal arches. The terminal sulcus of the tongue is derived from the ventral portion of the first pharyngeal pouch. The foramen caecum is located at the site where the interhyoid sulcus meets the first pharyngeal pouch. Pharyngeal arches 3, 4, and 5 fuse into the copula (hypobranchial eminence). **(156)** Dissected lower jaw with the tongue. Stage 7-5. The body of tongue is derived from the lateral lingual swellings and the impar tubercle. The lingual radix originates from the second pharyngeal (hyoid) arches. **(157)** Dissected tongue. Stage 8-2. Structures depicted: lingual body and radix; laryngeal opening delineated anteriorly by epiglottal swelling (derived from pharyngeal arches 3 and 4) and by arytenoid swellings (derived from pharyngeal arches 4 and 5). **(158)** Dissected oral cavity with tongue. Stage 8-2. Structures depicted: lower lip, gum, tongue (apex and corpus with papillae) epiglottal swelling, and entrance to pharynx.

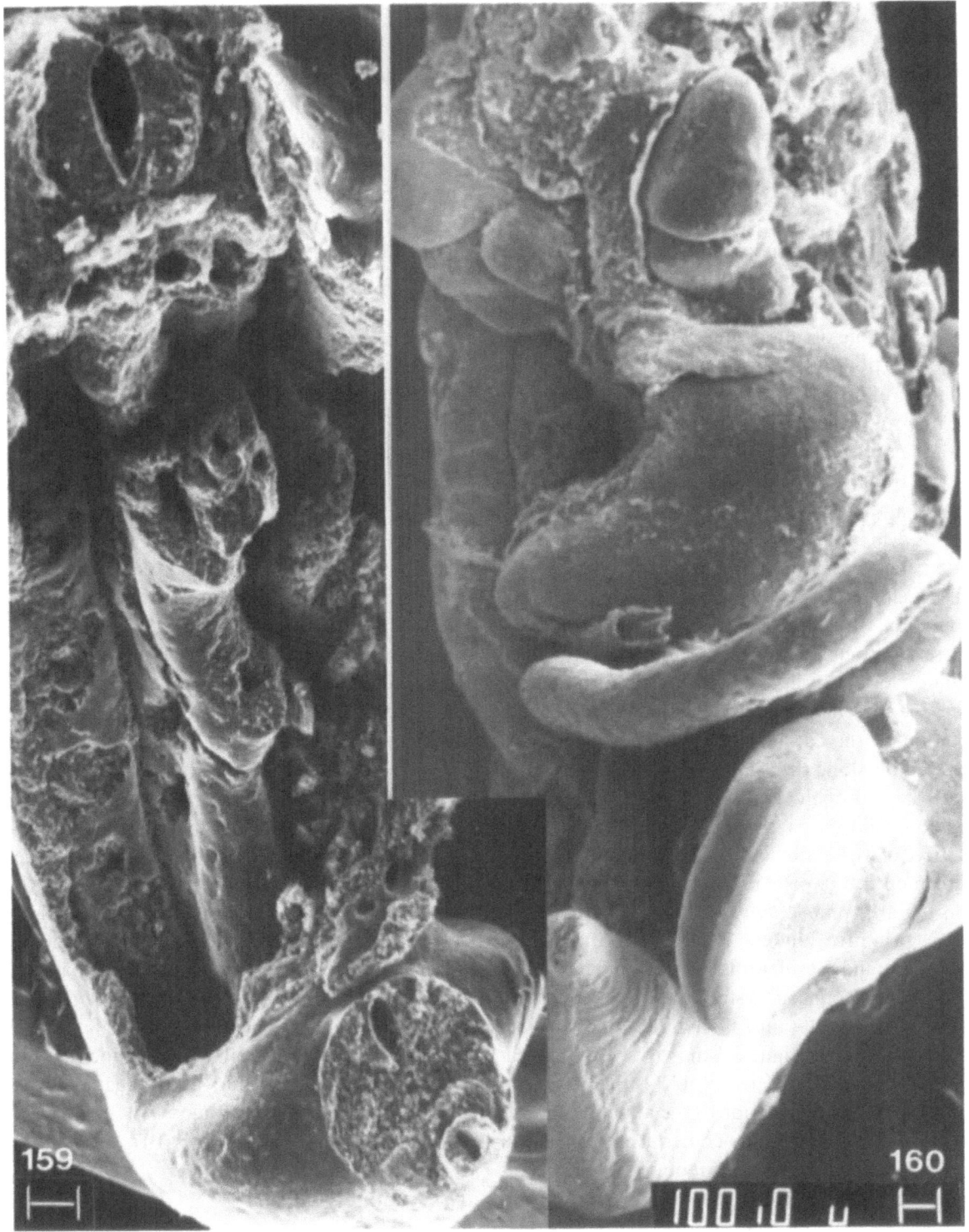

Figures 159 and 160. Digestive and respiratory organs. **(159)** Dissected coelomic cavity. Stage 7-2. Digestive tube suspended by dorsal mesentery in a longitudinal midline position. **(160)** Dissected visceral organs. Stage 7-4. Structures depicted: lungs, anterior mediastinum, stomach, gut loop, and rectum.

The intestines (Figs. 161–168)

The rapidly elongating midgut gives rise to the primary intestinal loop, which remains connected at the apex to the yolk sac by a long and narrow vitelline duct. In the axis of the loop, within the mesentery, is the superior mesenteric artery. A cranial limb and a caudal limb are distinguished on the primary intestinal loop. The cranial limb grows faster than the caudal limb and transforms into the distal portion of duodenum, the jejunum, and a major portion of the ileum. The caudal limb of the primary intestinal loop gives rise to the terminal portion of ileum and distends into the caecum with the appendix and into the proximal two-thirds of the transverse colon. The border between the cranial and caudal limbs of the primary interstinal loop is evident if a proximal portion of the vitelline duct persists as the Meckel's diverticulum.

The elongation of the gut is accompanied by formation of jejunoileal loops and by rotations. These rotations comprise a counterclockwise 270° rotation of the primary intestinal loop along the superior mesenteric artery and an additional 90° rotation of the caecum and ascending and transverse colon along the longitudinal axis of the body.

At the end of the embryonic period, the jejunoileal loop and the caecum extend into the proximal portion of the umbilical cord. At the end of the third month, the prolapsed intestinal loops return into the abdominal cavity.

The primitive hindgut is closed by the cloacal membrane and the allantois extends from its ventral wall into the connecting stalk. As the mesonephric ducts open laterally into the hindgut, the terminal portion of the hindgut is known as the cloaca. The transverse peritoneal fold – the urorectal septum – deepens posterior to the allantois toward the cloacal membrane, and consequently the cloaca becomes divided into a ventral portion – the primitive urogenital sinus with the allantois, and into a dorsal portion – the rectum.

The liver (Fig. 169)

The liver develops from an endodermal diverticulum of the terminal portion of the foregut, the future duodenum, into the ventral mesentery. From the dorsal portion of the hepatic diverticulum, trabecules of hepatic cells grow into the mesenchymal septum transversum containing vitelline (omphalomesenteric) veins. The hepatic trabecules intermingle with the blood sinusoid provided by vitelline veins and form hepatic lobules. The smaller portion of the hepatic diverticulum contributes to the gallbladder and the cystic duct. Hematopoetic tissue appears within the liver during the second month of development and disappears shortly before birth.

The pancreas

The pancreas originates from two endodermal (dorsal and ventral) evaginations from the duodenum. The dorsal evagination, which appears first, is located opposite and slightly cranial to the hepatic diverticulum and grows into the dorsal mesoduodenum. The ventral evagination grows into the ventral mesoduodenum close to the duodenal opening of the hepatic diverticulum. As the duodenum rotates, the ventral pancreatic primordium shifts dorsally and fuses with the dorsal primordium. Only a portion of the pancreatic head originates from the ventral primordium. The rest of the head, the corpus, and the cauda develop from the dorsal primordium. As the two portions fuse, the main ducts of both primordia join together. Finally the proximal portion of the main definitive pancreatic duct is provided by the ventral primordium, and its distal portion by the distal portion of the duct of the dorsal primordium. The proximal portion of the dorsal pancreatic duct persists in some individuals as the accessory pancreatic duct.

The definitive exocrine and endocrine pancreatic tissues originate from branching indifferent pancreatic tubules. The exocrine secretory tissue differentiates into the acini. The endocrine tissue detaches from the primitive pancreatic ducts as pancreatic islets of Langerhans.

The respiratory organs

The epiglottis and the larynx

The epiglottis develops behind the root of the tongue from the copula, which turns into the epiglottal swelling containing the mesenchyme of the third and fourth pharyngeal arches. The epiglottis is anterior to the laryngeal orifice, which is delineated by the arytenoid swellings. The arytenoid swellings originate from mesenchyme pro-

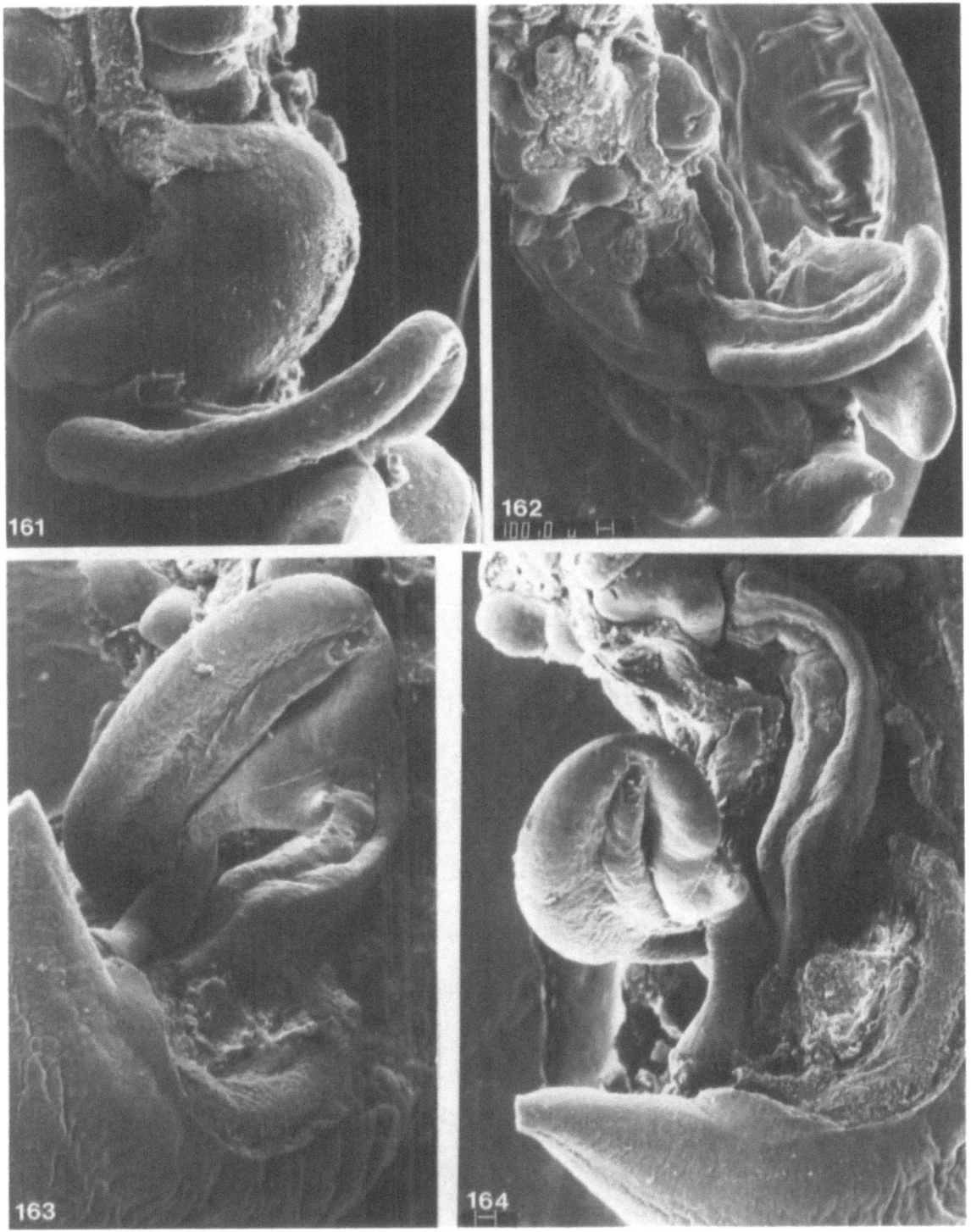

Figures 161–164. Gut loop. Stage 7-4. **(161)** Anterior mediastinum, stomach, and gut loop. Frontal view. **(162)** Exposed gut loop rotating around the superior mesenteric artery (stomach removed). **(163)** Gut loop rotating around the superior mesenteric artery. The caudal limb of the loop exhibits caecal dilatation. Distinct taenia is present on the primitive colon. Inferior view. **(164)** Gut loop (left lateral view). Inferior limb with caecal dilatation, appendix, colon, and rectum. Lateral to the gut are gonadal and mesonephric ridges.

92

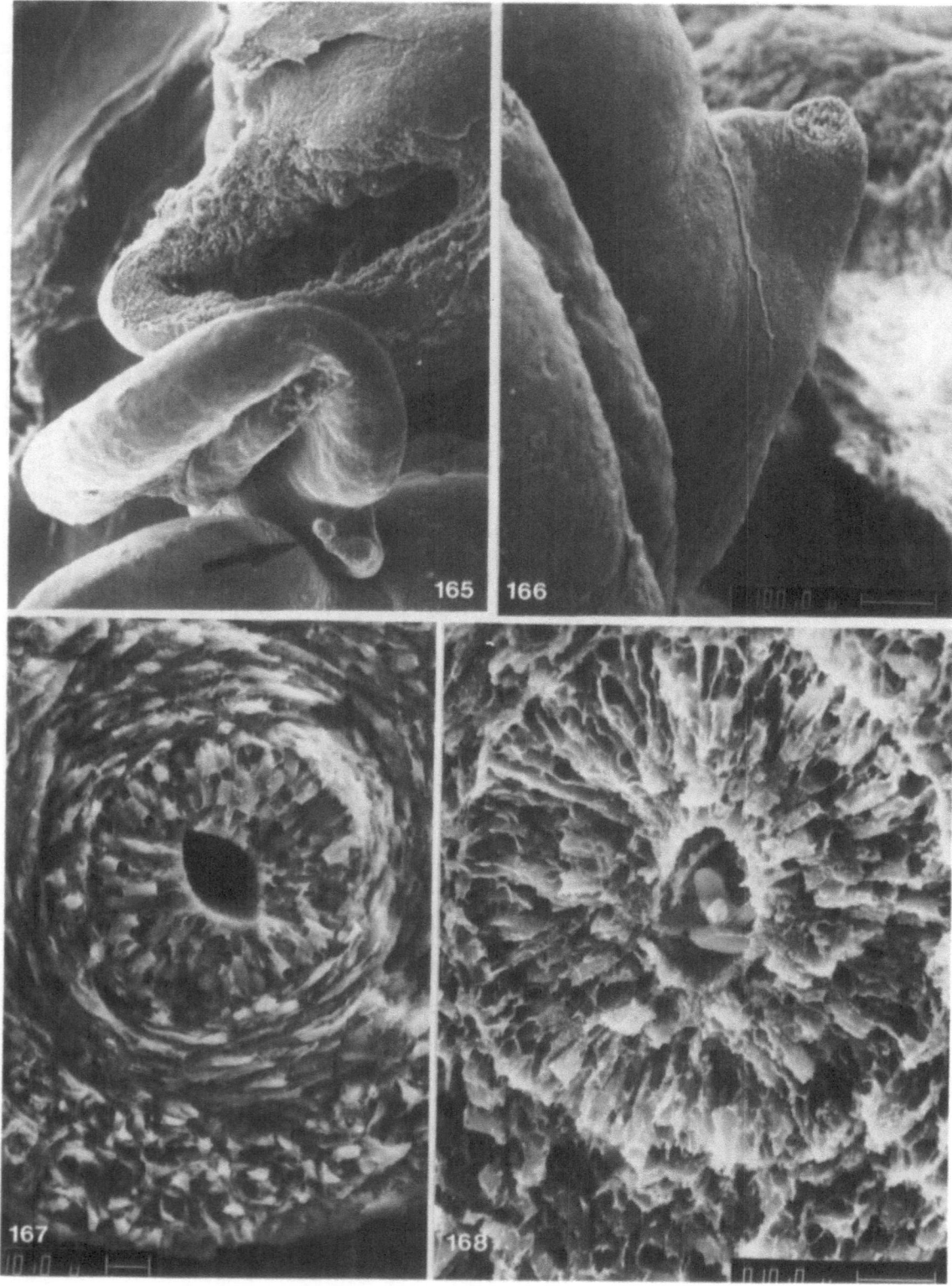

Figures 165–168. (165) Dissected stomach and primitive gut loop. The caecal bud with the appendix (arrow) are evident on the distal limb of the gut loop. **(166)** Caecal swelling of the gut with a characteristic taenia. **(167)** Transverse section of a primitive gut. Stage 8-1. The lumen is lined by an endodermal stratified cylindrical epithelium containing glycogen vacuoles. Mesenchymal cells form an inner circular layer and an outer longitudinal layer. **(168)** Primitieve stratified endodermal epithelium of the embryonic duodenal loop.

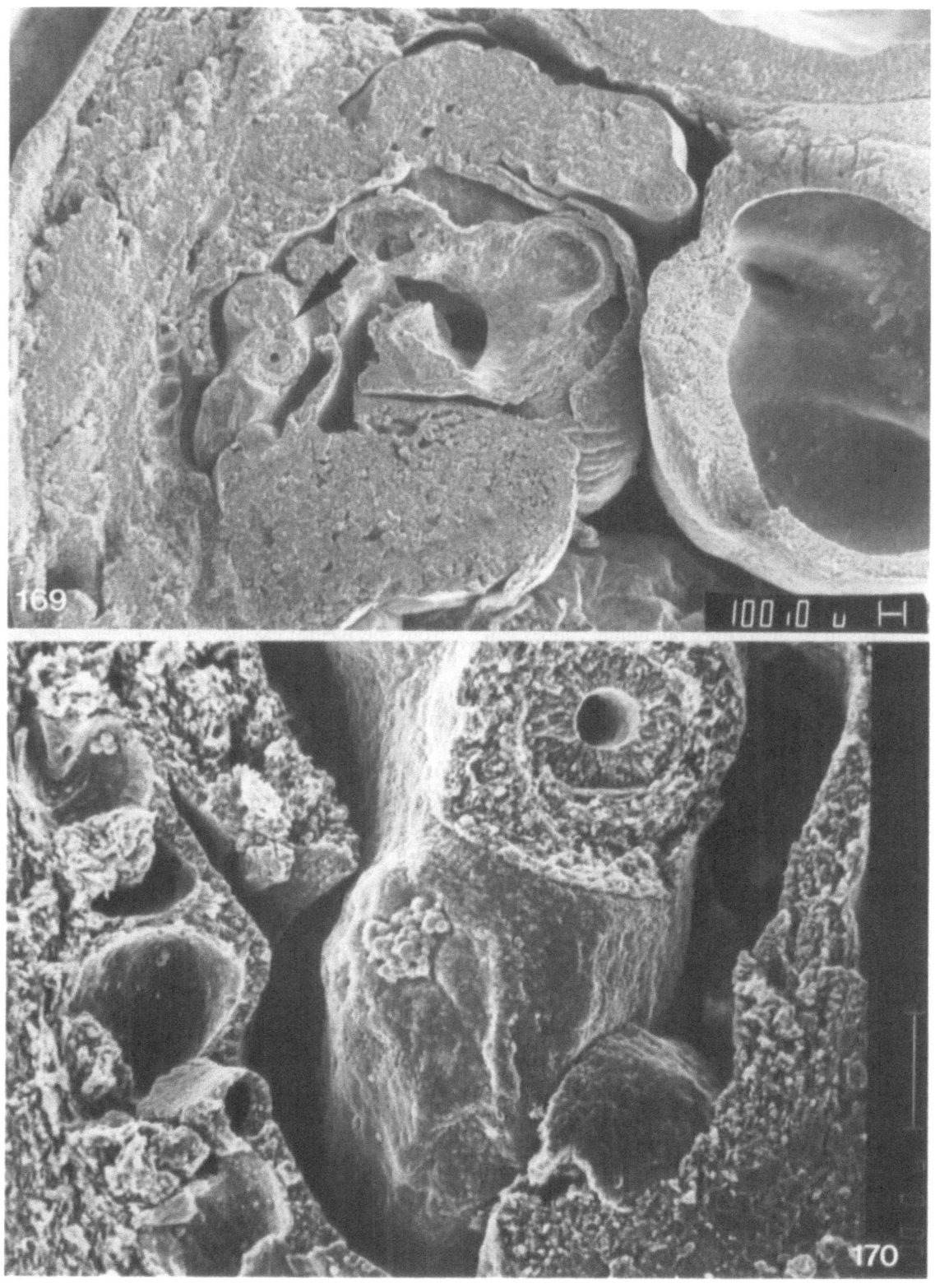

Figures 169 and 170. (169) Sagittal dissection of head, neck, and thorax. Stage 7-2. Structures depicted: oral cavity, left lung (arrow) located within the pleural cavity, and liver. **(170)** Detail of the left lung with a dissected main bronchus.

vided by the fifth pharyngeal arches. As the primordium of the tracheobronchal tree becomes detached from the esophagus, the larynx becomes evident. As the trachea grows, the upper portion of the larynx is pushed into the pharyngeal cavity.

The trachea and the lungs (Figs. 79C, 80C, and 169–177)

A portion of the (endodermal) foregut caudal to the pharynx is known as the primitive esophagus. At the border between pharynx and esophagus, a respiratory diverticulum from the ventral wall of the esophagus detaches by longitudinal esophageotracheal furrows. As the furrows deepen and fuse, the primitive larynx and trachea are separated from the esophagus The respiratory diverticulum branches into two bronchi, which together with the lung mesenchyme, constitute the lung buds. The bronchus of the right lung divides into three main bronchi and the left bronchus into two main bronchi of the proper pulmomary lobes. The bronchi divide dichotomically. The lungs grow into the coelomic (or pleuropericardioperitoneal) cavity dorsal to the heart.

Those portions of the pleuropericardioperitoneal cavity that are occupied by the developing lungs represent the primitive pleural cavities. The respiratory units of the pulmonary lobules become evident during the fifth month. The terminal bronchioli split into several alveolar ducts from which the alveoli expand as the breathing begins. The development of the alveoli is related to the arrangement of pulmonary vessels. The pulmonary arteries accompanying the the bronchioli split into capillaries oriented at right angles to the alveolar duct. The capillaries enter veins located between pulmonary lobules. The alveoli expand into the space left between the capillaries and the veins. The expansion of pulmonary alveoli and the difusion of oxygen from the alveoli into the blood circulating in capillaries are important factors limiting extrauterine fetal survival.

The coelom and the mesenteries

The coelem

The term coelom is applied to a cavity lined by the mesoderm. The coelomic cavity is located between the somatic mesoderm layer and the splanchnic mesoderm layer. The extraembryonic coelom appears within the primary mesoderm of the chorionic sac. The extraembryonic coelom obliterates as the amnion expands. During this process, the mesoderm layer of the amnion fuses with the mesoderm layer of the chorion and of the yolk sac. Coincidently the umbilical cord is formed.

The intraembryonic coelom originates from fusion of the pericardial cavity (developing separately around the heart tube) with the peritoneal cavity (appearing in the lateral mesoderm, lateroventral to the somites). In the early embryos, the pericardial cavity communicates with the peritoneal cavity by two pericardioperitoneal canals located along the broad dorsal mesentery of the foregut. As the developing lungs evaginate, mostly into the pericardioperitoneal canals, the pericardioperitoneal canals represent primordia of the pleural cavities. The pericardial cavity becomes separated from the pleural cavities by two pleuropericardial folds containing common cardinal veins and the phrenic nerves. The pleuropericardial folds fuse into the pleuropericardial membrane, the primordium of the fibrous pericardium.

The pleural cavities are separated from the peritoneal cavity by the diaphragm. The ventral portion of the diaphragm develops from the septum transversum, the dorsal portion from the pleuroperitoneal folds. The septum transversum is a mesenchymal plate descending with the heart, located between the heart, the stalk of the yolk sac, and the hepatic diverticulum. The pleuroperitoneal folds, which are related to the cranial portions of mesonephros, obturate the pericardioperitoneal canals separating the pleural cavities from the peritoneal cavities.

The mesenteries

The mesentery, a duplicature of splanchnic mesoderm, attaches the digestive tube to the body wall. Both mesenteries, the ventral and the dorsal, are present in the abdominal portion of the esophagus, in the stomach, and in the upper portion of duodenum (including the liver). The lower portion of the duodenum, the jejunum, the ileum, the colon, the caecum including the appendix, and the sigmoideum are suspended only on the dorsal mesentery connecting the gut with the dorsal body wall.

As the liver develops within the ventral mesentery, the ventral portion of the ventral mesentery located between the anterior body wall and the liver becomes the falciform ligament of the liver. The dorsal portion of the ventral mesentery located between the liver and the stomach contributes to the lesser omentum. Its free margin containing the common bile duct, the portal vein, and the hepatic artery delineates the upper portion of the epiploic foramen of Winslow.

The dorsal mesentery of the stomach and the upper duodenum contains the primordium of the spleen in an intraperitoneal localization and contributes to the greater omentum. In relation to the positional changes of the stomach and liver, the hepatogastric and hepatoduodenal portions of the lesser omentum change into the lining of the lesser peritoneal sac. The greater omentum forms a duplicature – the greater peritoneal sac – extending over the intestinal loops. The greater omentum attaches to the mesentery of the transverse colon. The cavity in this duplicature obliterates postnatally. As the gut rotates, the mesentery twists around the superior mesenteric artery. As the duodenum and the colon attain their final position, the mesentery of the duodenum and the colon ascendens and descendens are pressed to the abdominal wall. The apposed mesentery fuses with the somatic mesoderm layer (the peritoneum). Consequently, the duodenum and the colon ascendens and descendens become located retroperitoneally.

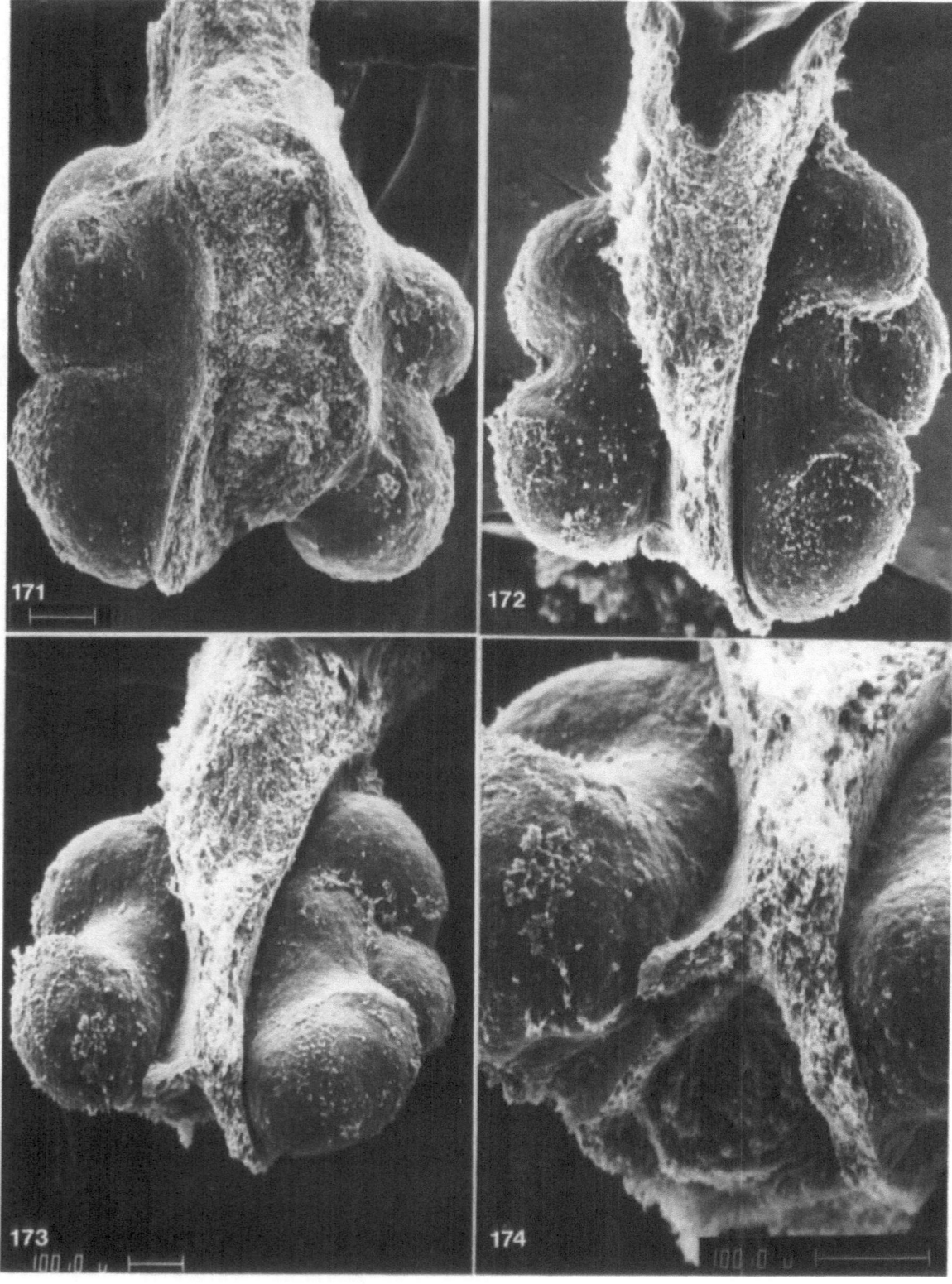

Figures 171–174. Dissected lungs. Stage 7-4. **(171)** The right lung consists of three lobes, the left of two lobes. Dissected anterior mediastinum. Anterior view. **(172)** Posterior view of the specimen. Dissected posterior mediastinum. **(173)** Posteroinferior view of the specimen. **(174)** Bifurcation of the posterior mediastinum containing the esophagus.

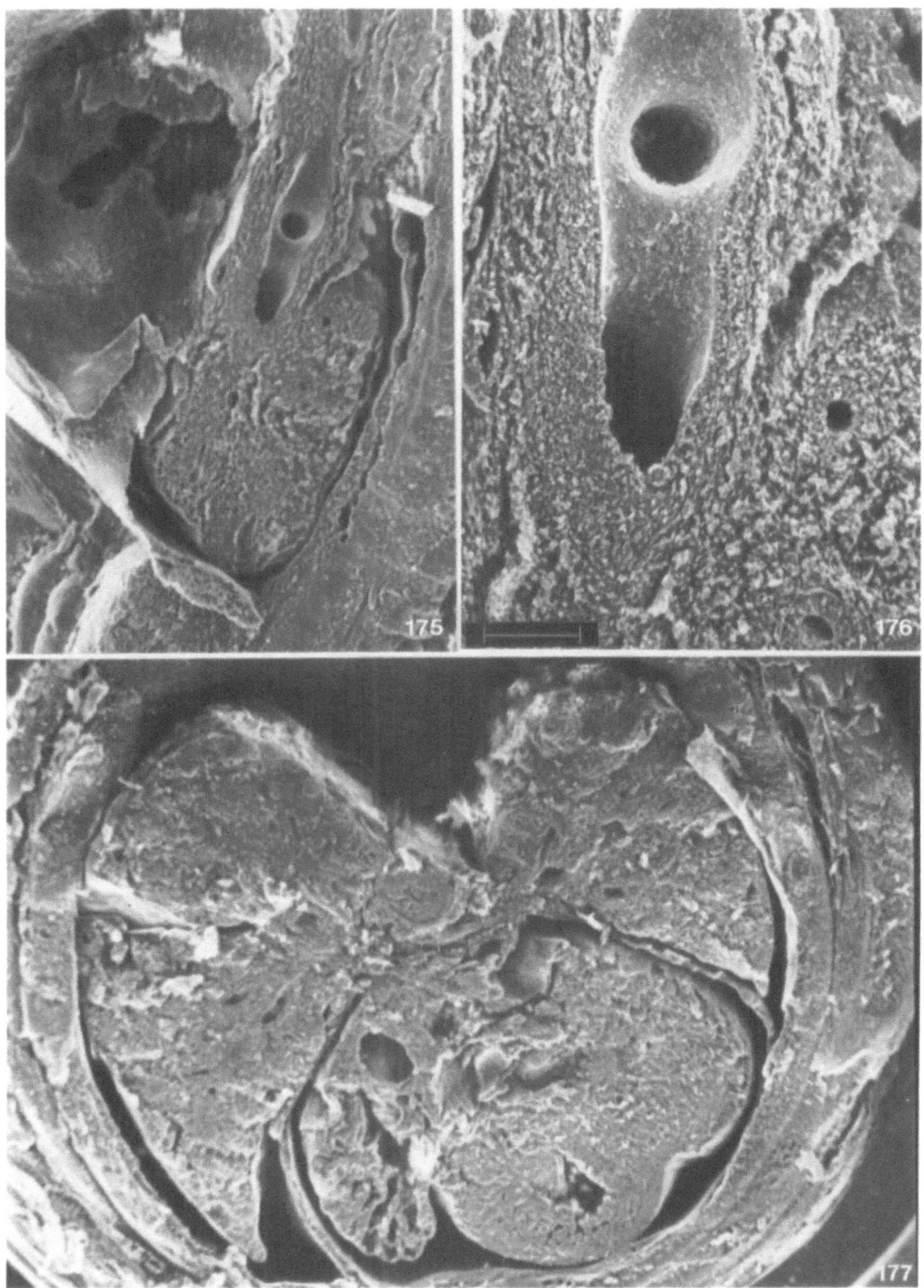

Figures 175–177. (175) Lungs within the pleural cavity. Stage 8-1. **(176)** Detail of a dividing bronchus. **(177)** Transverse dissection of thorax. Stage 8-2. Lungs located within the pleural cavity; heart within the pericardial cavity; the pericardial cavity separated from the pleural cavity by the pleuropericardial membranes.

6. ORGANS WITH ECTODERMAL COMPONENTS

The skin and epidermal derivatives (Figs. 178–180)

The epidermis is an epithelium originating from the surface ectoderm. Nails, hair, sebaceous glands, sweat glands, and mammary glands are epidermal derivatives. The dermis (corium) is mesenchymal in origin.

The *skin* develops from the single-layered cuboidal epithelium of the surface ectoderm, which changes into the embryonic epidermis represented by a bilaminar cuboidal epithelium with a basal layer and a superficial peridermal layer (periderm, epitrichium). During the third month, an intermediate layer is formed and consequently the epidermis changes into a multilayered squamous epithelium. The mesenchyme underneath the surface epithelium changes into the dermis.

The *fingerprints* consist of dermal connective tissue ridges that are invaded by epidermal interpapillary pegs with primordia of sweat glands. They appear during the fetal months 3 and 4.

The *nails* appear in the third month as areas of thickened cornified epidermis.

The *hair* is formed from ectodermal epidermal buds growing into the underlying mesenchyme (dermis). The first generation of fetal hair is called lanugo.

The *sebaceous glands* originate from ectodermal buds growing from the external root sheath of the hair follicles. The epidermis of a newborn is covered by the vernix caseosa, a mixture of desquamate cornified cells of epidermis, hair, and secretory products of the sebaceous glands.

The *sweat glands* are formed during fetal months 5 and 6 as simple epithelial cords growing from the epidermis into the dermis.

The *mammary glands* are derived from the thoracic portion of the mammary ridge. The mammary ridges are linear ectodermal thickenings extending from the axillary to the inguinal areas. The ridges regress except for a discoid primordium (bud) of the future nipples. From the nipple grow four or five branching main mammary cords corresponding to the mammary lobes. At the end of intrauterine life, the epithelium of the nipple disintegrates, the nipple becomes everted, and 16–25 main mammary ducts gain proper openings.

The neuroectodermal organs

The central nervous system (Figs. 182–187)
Neural tissue originates from the neuroectoderm. In trilaminar presomite embryos, the neuroectoderm constitutes a neural plate with a broad anterior portion, the primordium of the brain, and a narrower posterior portion, the primordium of the medulla. The neural plate changes into a neural groove with elevated neural folds. The neural folds approach each other in the midline and their dorsal portions fuse, constituting the neural tube. The fusion of neural folds begins in the area of the fourth somite in embryos with seven paired somites and extends cranially as well as caudally. The temporarily anterior opening of the neural tube is known as the anterior neuropore, the posterior opening as the posterior neuropore. The anterior neuropore closes in 20-somite embryos and the posterior closes in 25- to 30-somite embryos.

The brain (Figs. 81c–86c and 188–207)
Primordia of three primary brain vesicles – the prosencephalon (forebrain), the mesencephalon (midbrain), and the rhombencephalon (hindbrain) – are recognized prior to the closure of the neural tube. The rapid growth of the neural tissue in somite embryos and the ventral apposition of the neural tube to the blastematous vertebrae account

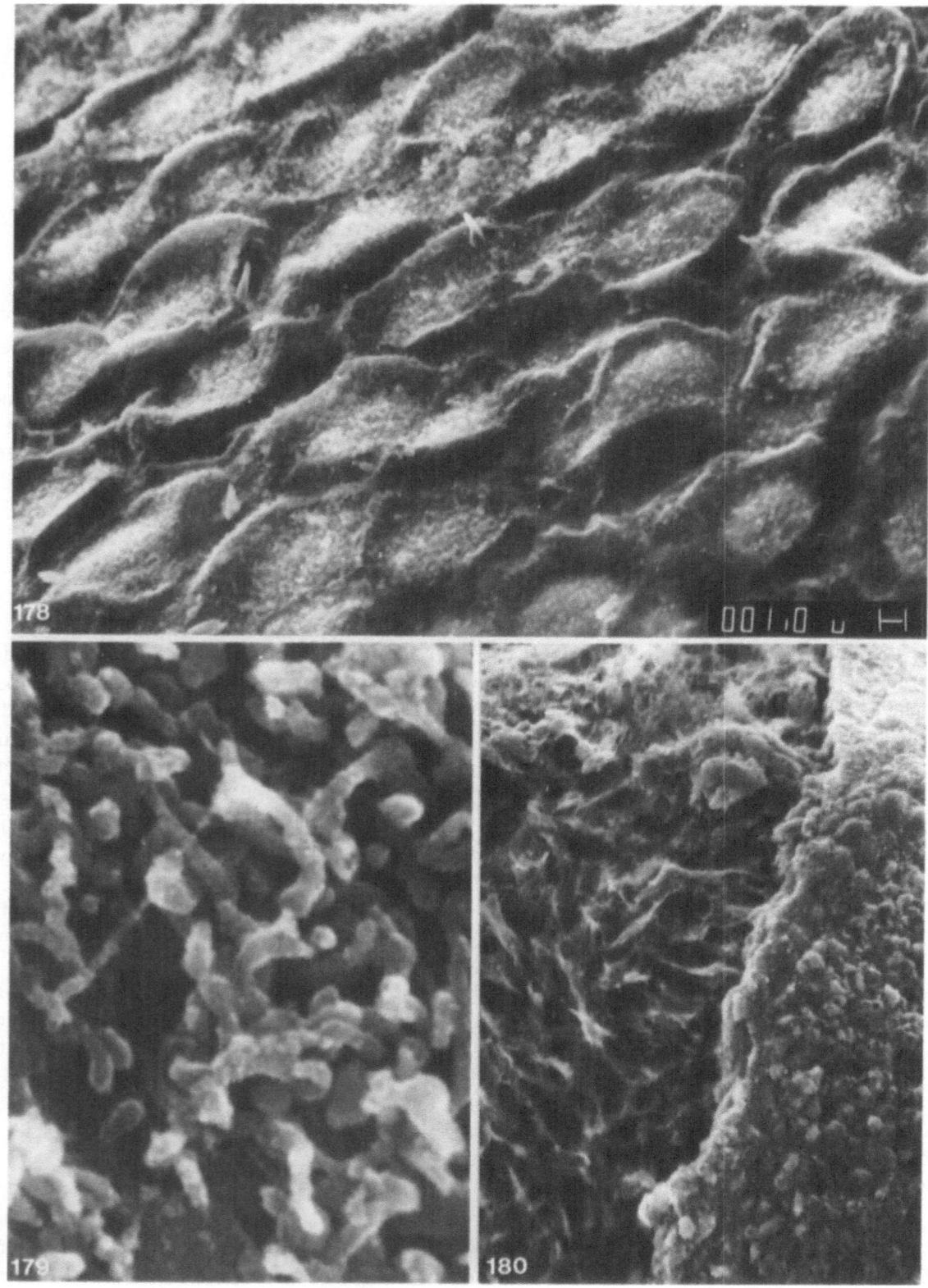

Figures 178–180. The skin. **(178)** Peridermal cells on the surface of primitive epidermis. **(179)** Microvilli on the surface of peridermal cells. **(180)** Primitive skin composed of surface ectoderm (on the right) and underlying mesenchyme (on the left).

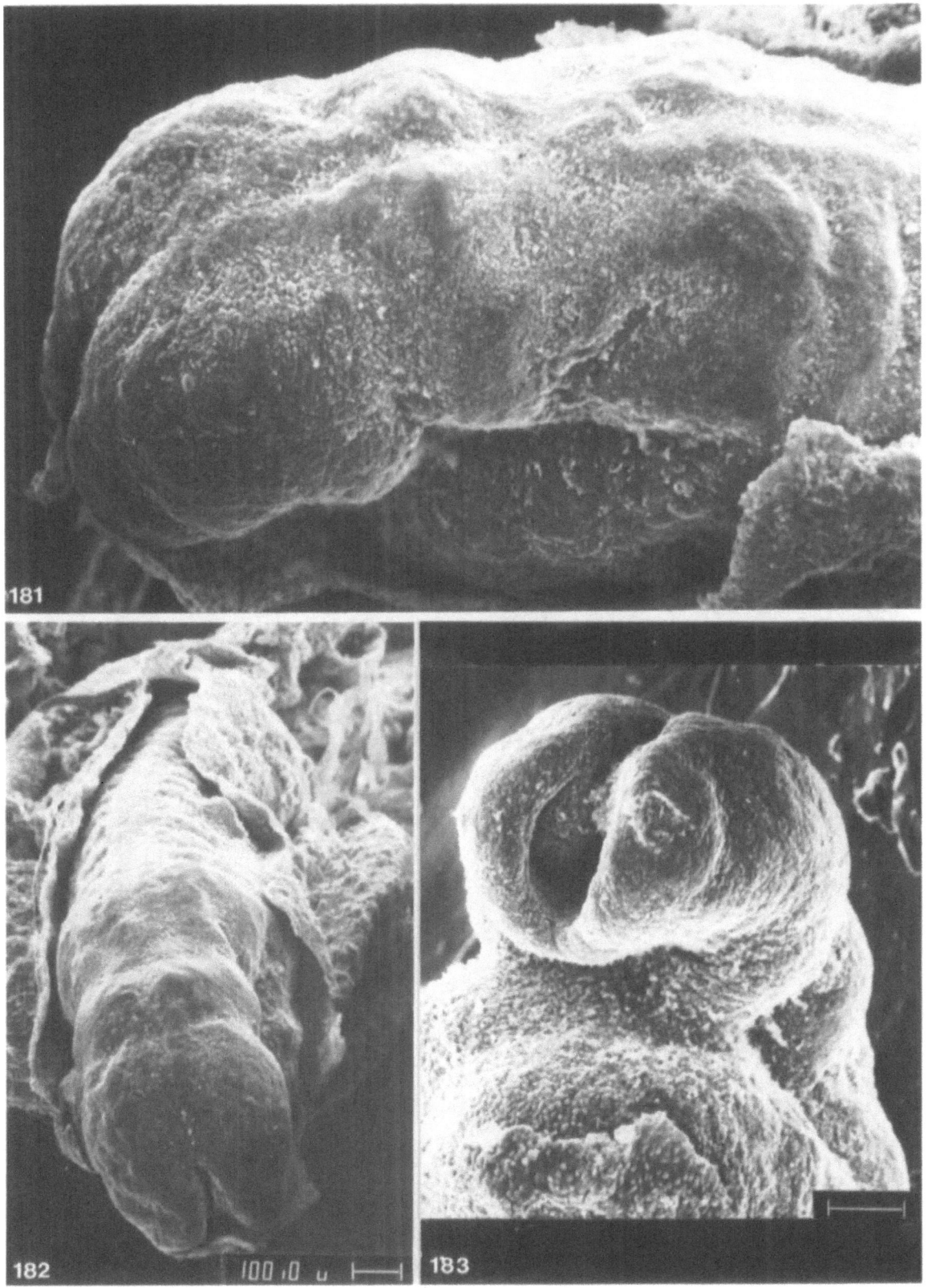

Figures 181–183. Primary cerebral vesicles. Stage 6-2. **(181)** Embryonic head with primordia of prosencephalon, mesencephalon, and rhombencephalon. Left posterolateral view. **(182)** Prosencephalon with an open anterior neuropore, closed mesencephalon, rhombencephalon, and medullary tube. **(183)** Anterior neuropore. Left anterior view.

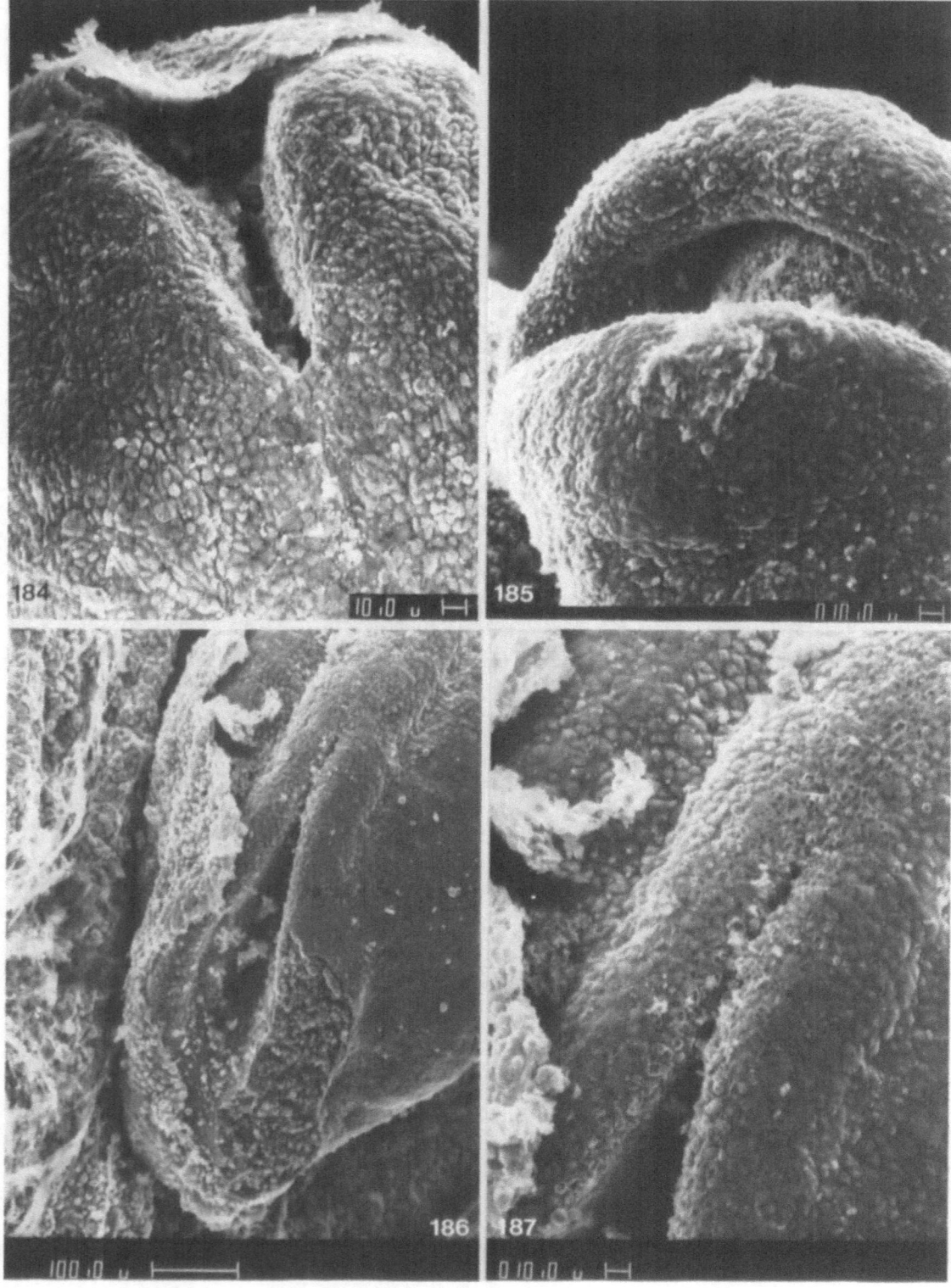

Figures 184–187. Closing neural tube. Stage 6-2. **(184)** Anterior neuropore covered partially by a dissected amnion. Dorsal view. **(185)** Open prosencephalic vesicle; 'evagination' of the optic cup. **(186)** Posterior neuropore. **(187)** Closing posterior neuropore. Note the 'bubble cells' on the rims.

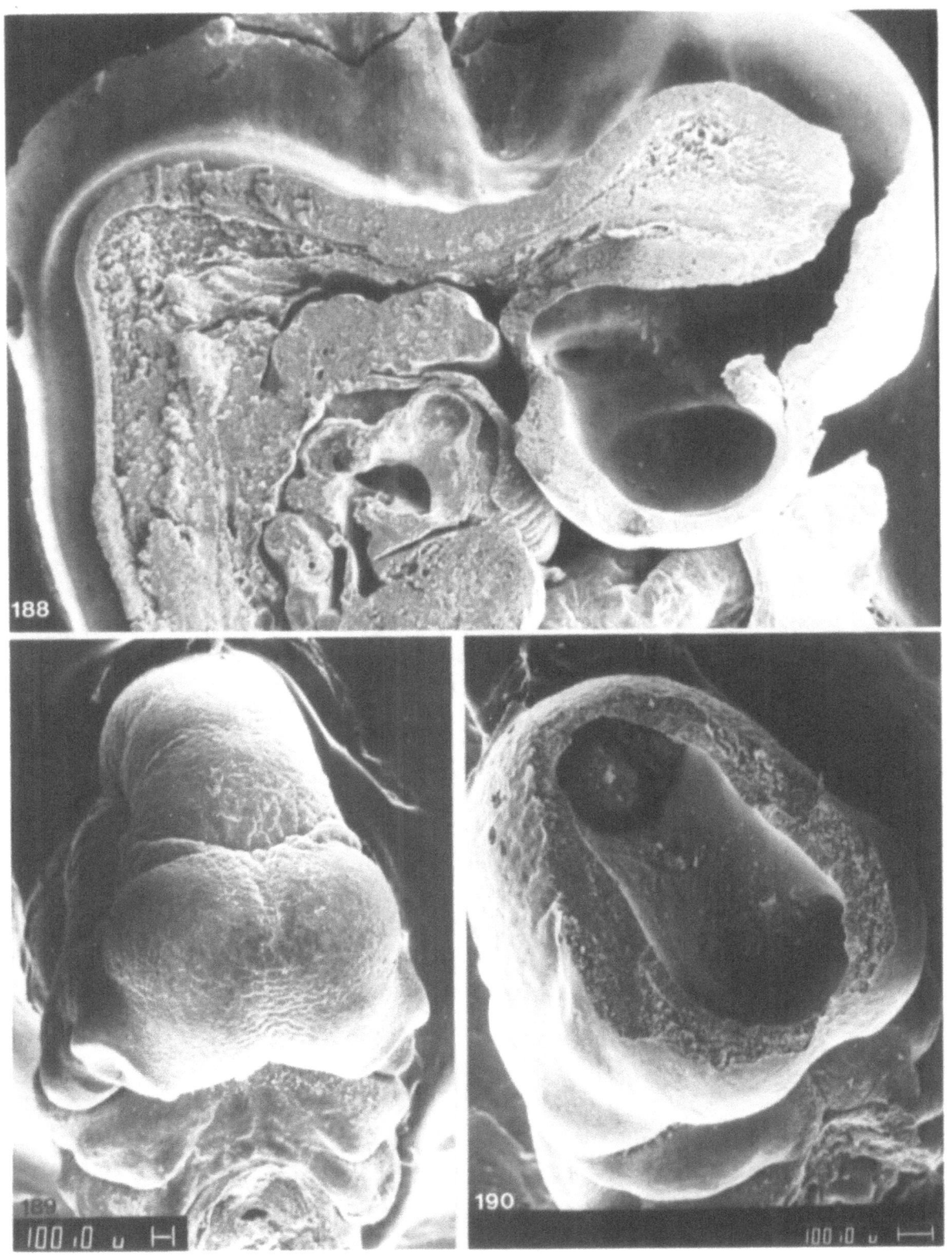

Figures 188–190. (188) Sagittally dissected embryo. Stage 7–4. CNS structures depicted: telencephalon with an evaginating hemisphere, diencephalon with opening of the stalk of optic cup, mesencephalon, fourth ventricle, and metencephalon and myelencephalon with medullary tube. Cephalic, cervical, and pontine flexures. **(189)** Head of an embryo. Stage 7-3. CNS structures depicted: hemispheres and mesencephalon. **(190)** Dissected brain. Stage 7-3.

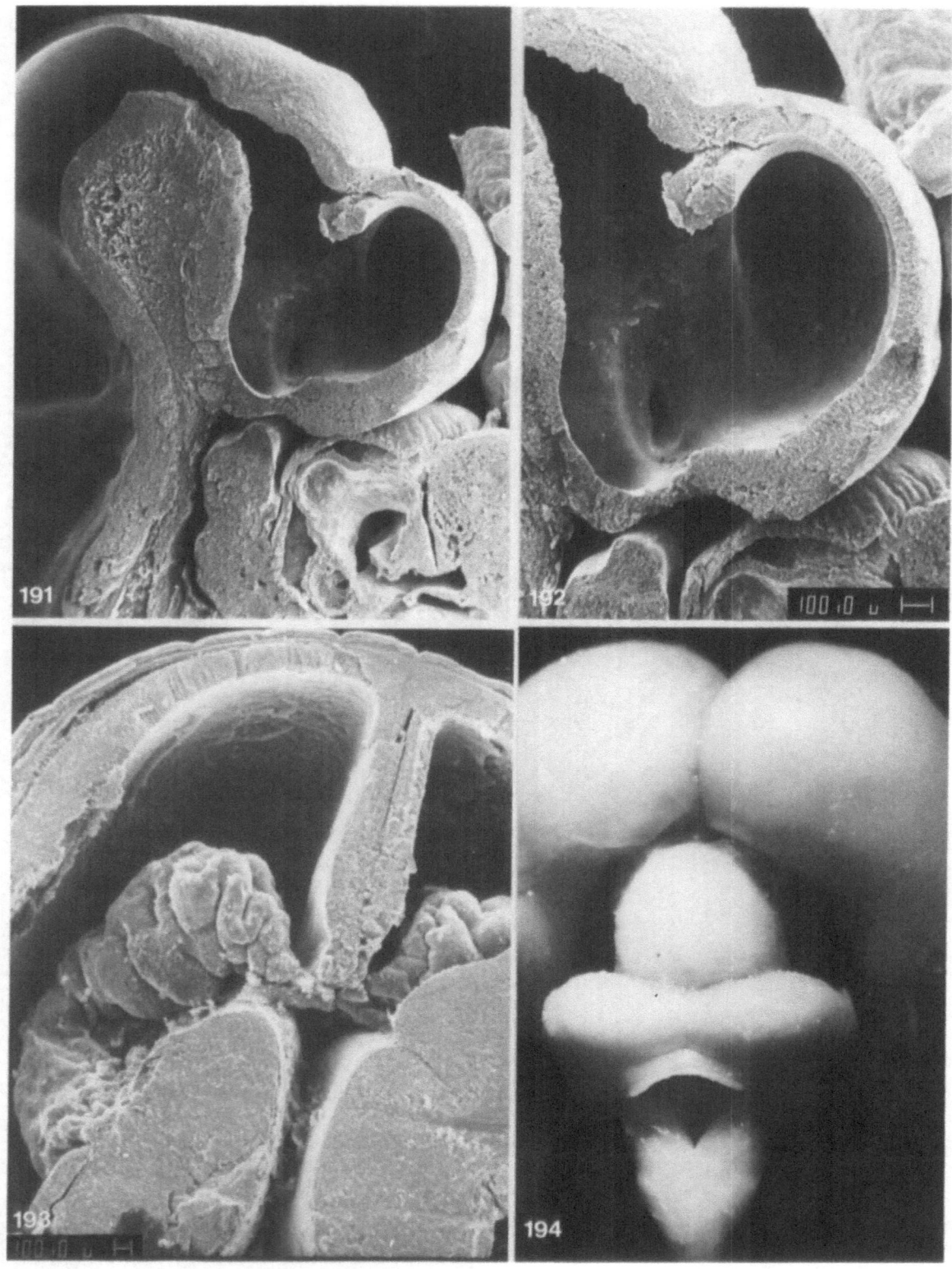

Figures 191–194. (191) Dissected brain. Stage 7-4. Structures depicted: left hemisphere with anterior commissure, broad inter-ventricular foramen, diencephalon with opening of the stalk of optic cup, mesencephalon, and fourth ventricle. **(192)** Telencephalon and diencephalon. Stage 7-4 **(193)** Choroid plexus protruding into the lateral ventricle of the hemispheres. Stage 8-1 **(193)** Dorsal view of fetal brain (fourth gestational month). Structures depicted: cerebral hemispheres, mesencephalon, cerebellar hemispheres with an indistinct vermis, posterior medullary velum, cut edge of the roof of fourth ventricle, and medulla oblongata.

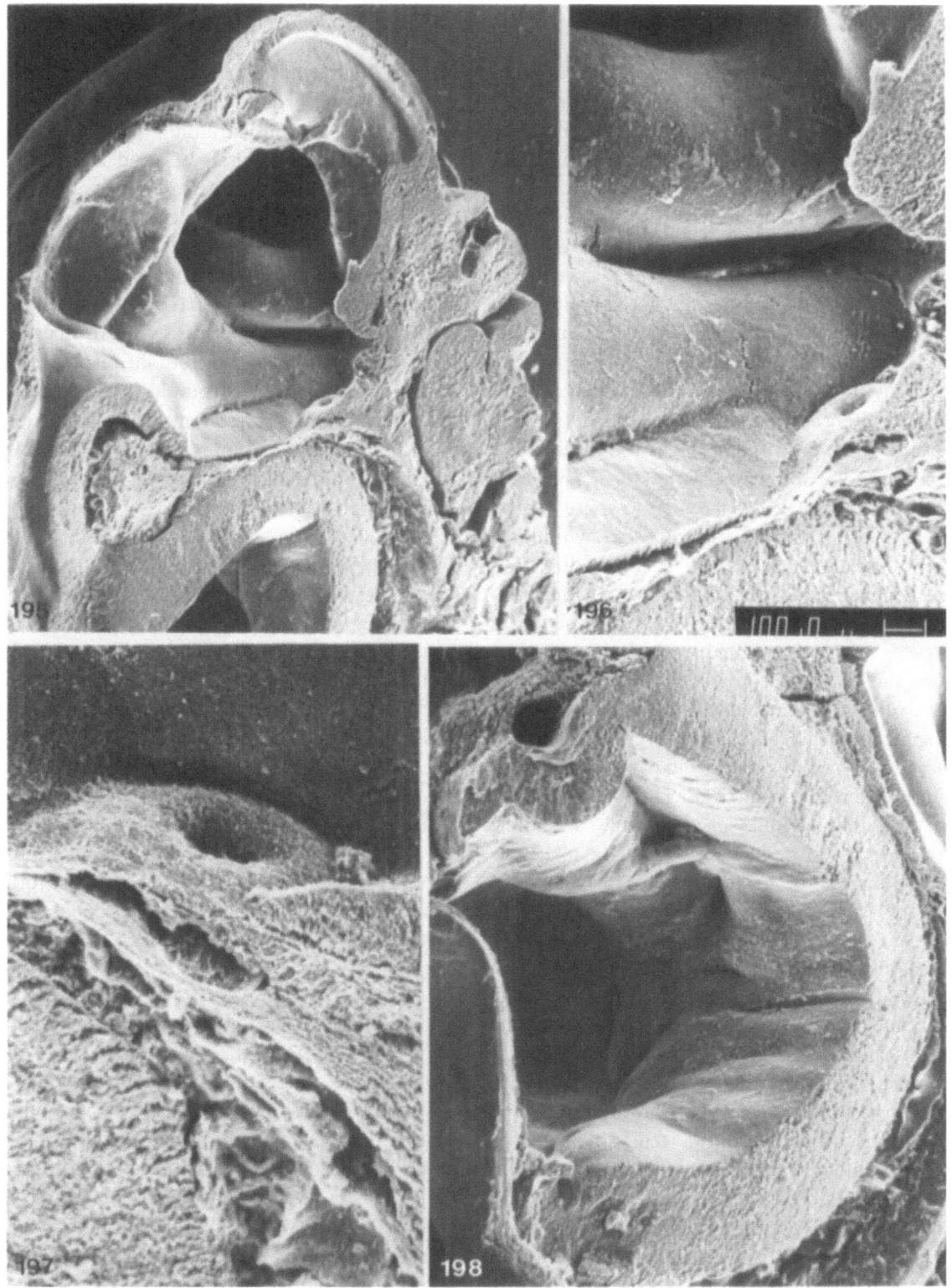

Figures 195–189. (195) Parasagittally dissected embryonic head. Stage 8-1. Structures depicted: parasagittally dissected hemisphere, broad interventricular foramen, corpus striatum, hypothalamus with the infundibulum of neurohypophysis, thalamus, mesencephalon, and dissected fourth ventricle. **(196)** Basal and lateral portions of the hypothalamus with the hypophyseal infundibulum. **(197)** Hypophyseal infundibulum. **(198)** Dissected fourth ventricle: rhomboid fossa, pontine flexure, medulla oblongata with four distinct neuromeres, cut rhomboic lip (cerebellar primordium), primordium of choroid plexus, and cut inferior medullary velum.

for the ventral bending of the CNS. A cephalic flexure develops in the mesencephalic area, and a cervical flexure is formed between the hindbrain and the medullary tube. Both these flexures are dorsally convex. A third flexure, dorsally concave, which becomes evident in the area of the hindbrain, is known as the pontine flexure. The forebrain (prosencephalon) subdivides into the telencephalon (endbrain) and the diencephalon; the hindbrain (rhombencephalon) subdivides into the metencephalon and the myelencephalon.

The cavity of the brain vesicles, which is filled with the cerebrospinal fluid, becomes subdivided accordin to the cleavage of the brain vesicles. The lateral ventricles are present within the hemispheres communicating each by one interventricular foramen with the third ventricle located within the impar telencephalon and the diencephalon. The mesencephalic aqueduct, present within mesencephalon, connects the third ventricle with the fourth ventricle, which is located within the metencephalon and the myelencephalon. The cavity of the fourth ventricle extends into the medullary tube as the spinal central canal. The cerebrospinal fluid (liquor) is formed by the choroid plexuses evaginating into the cerebral ventricles. The fluid is drained into the subarachnoid space through a midline centrally located foramen and two lateral apertures in the ceiling of the fourth ventricle.

The telencephalon. The telencephalon consists of a median unpaired telencephalon impar and two lateral evaginations, the cerebral hemispheres. The development of the hemispheres is related to the olfactory placodes. The primitive hemispheres expand over the diencephalon and mesencephalon, each of them differentiating into a frontal, a parietal, an occipital, and a temporal lobe. A depressed lateral area between the frontal and temporal lobes is known as the insula, which becomes covered by the enlarging temporal lobe. The surface of the hemispheres is originally smooth. Gyri and sulci appear during the second half of intrauterine life.

A medial, a basal, and a lateral wall can be distinguished on each hemisphere. Vascularized connective tissue (covered by the ependymal lining of the brain vesicles) penetrates the cavity of the two lateral ventricles in an area known as the choroid fissure and forms the choroid plexus of the lateral ventricle. The thick basal wall of the hemisphere is the corpus striatum, which fuses with diencephalon. The lateral wall and a part of the medial wall of each hemisphere represent the pallium, the primordium of the cerebral cortex.

The anterior wall of the telencephalon impar is the terminal lamina. The corpus callosum develops in the superoposterior portion of the terminal lamina and is contributed to by nerve fibers connecting cortical areas of both hemispheres. The corpus callosum extends from the lamina terminalis over the roof of the third ventricle. The portion of the lamina terminalis anterior to and underneath the corpus callosum changes into the septum pellucidum.

The diencephalon. The diencephalon consists of a roof plate, two lateral walls, and a ventral plate. The cavity within the diencephalon is the third ventricle. The most caudal part of the diencephalic roof evaginates as the primordium of the pineal gland. The area close to the pineal gland represents the epithalamus. The lateral wall of diencephalon is devided by a hypothalamic sulcus into the thalamus located dorsally and the hypothalamus located ventrally. The ventral plate of hypothalamus forms the tuber cinereum and evaginates the neurohypophyseal primordium. Optic evaginations are formed from the anterior portion of the diencephalon from an area where consquently the optic chiasma is located.

The mesencephalon. The mesencephalon exhibits the basic structure of the neural tube. Its cavity is known as the mesencephalic canal (aqueduct of Sylvius), which shows a distinct sulcus limitans separating the alar plate located dorsally from the ventrally located basal plate. The mesencephalic alar plate is known as the tectum. As on the surface of the tectum two primary longitudinal ridges become subdivided by a transverse furrow, the corpora quadrigemina, with two anterior and two posterior collicles, become evident.

The metencephalon (cerebellum and pons) and myelencephalon (medulla oblongata). The development of metencephalon and that of the myelencephalon are closely related. Both of them develop on

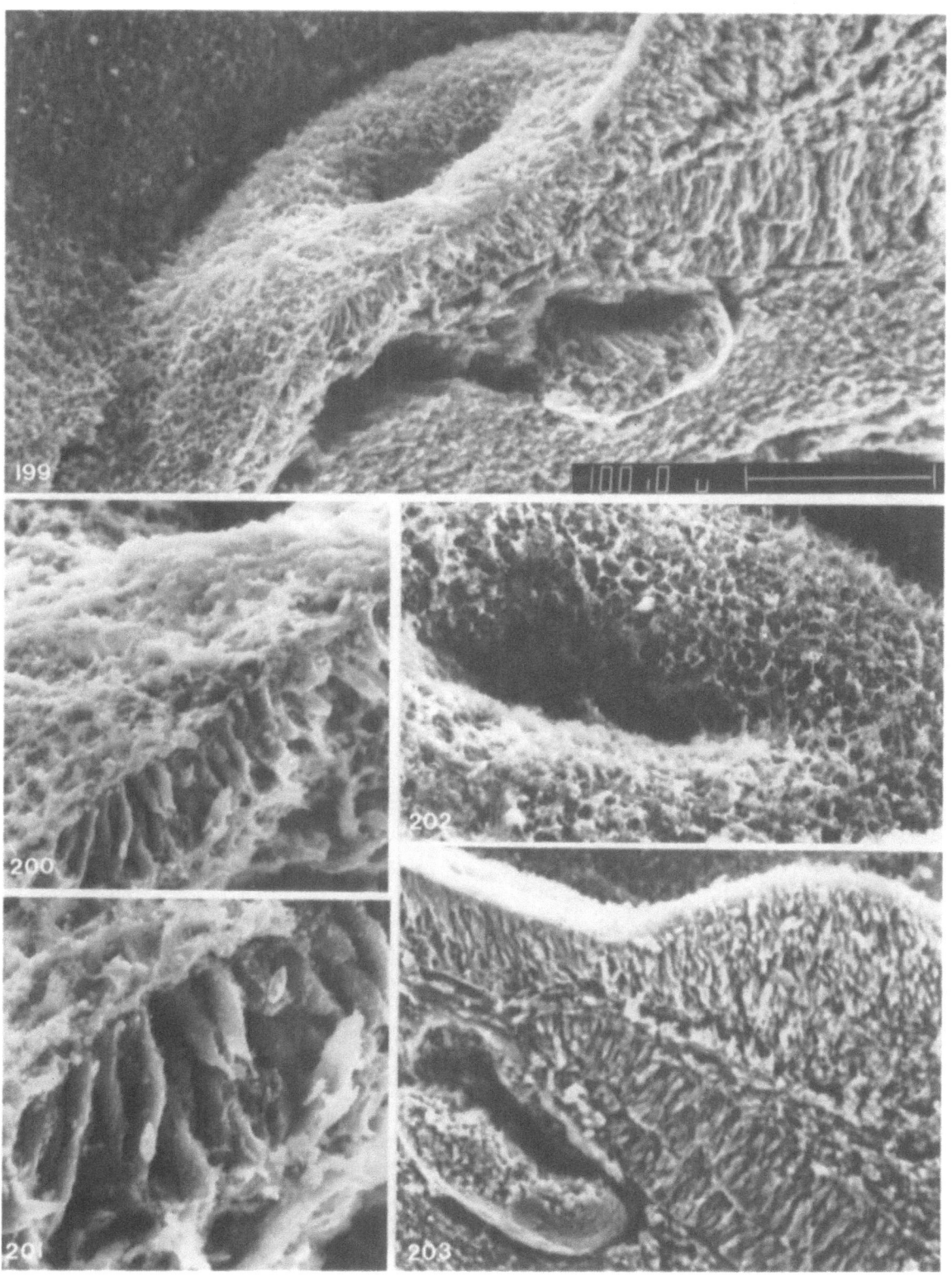

Figures 199–203. Hypophyseal primordium. Stage 8-1 **(199)** Hypophyseal infundibulum on the floor of the third ventricle. Parasagittal dissection. **(200 and 201)** Special spongioblasts (possibly tanycytes) near hypophyseal infundibulum. **(202)** Surface of hypophyseal infundibulum. **(203)** Parasagittally dissected infundibulum with contacting Rathke's pouch.

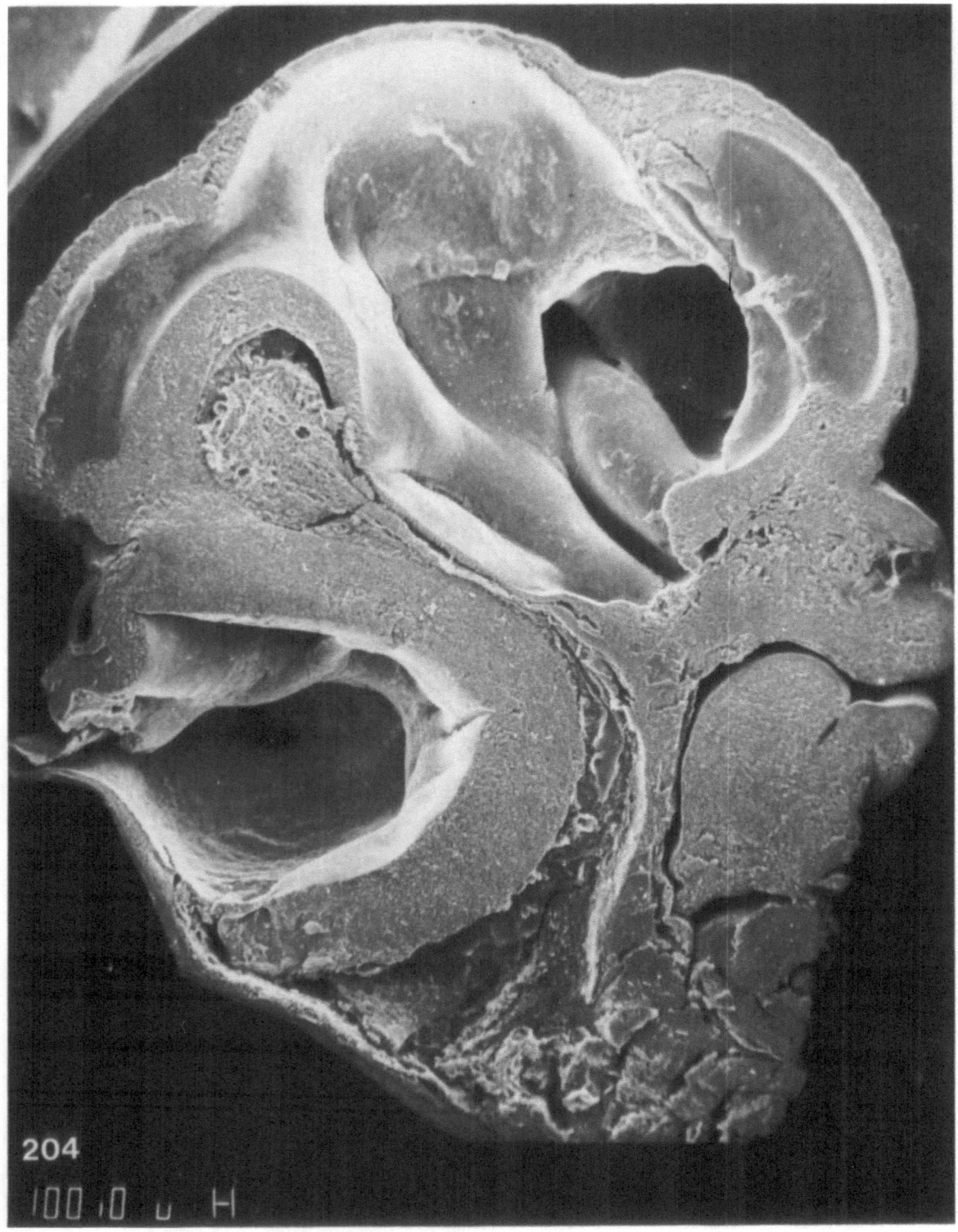

Figure 204. Parasagittally dissected embryonic head and neck. Stage 8-1. Structures depicted: nose, oral cavity with the tongue, and pharyngeal cavity. Components of the CNS: cerebral hemispheres, corpus striatum, lamina terminalis, interventricular foramen, epithalamus, thalamus, hypothalamus, mesencephalic floor (tegmentum), cavity of mesencephalon (note the mesencephalic flexure), and fourth ventricle (note the pontine flexure).

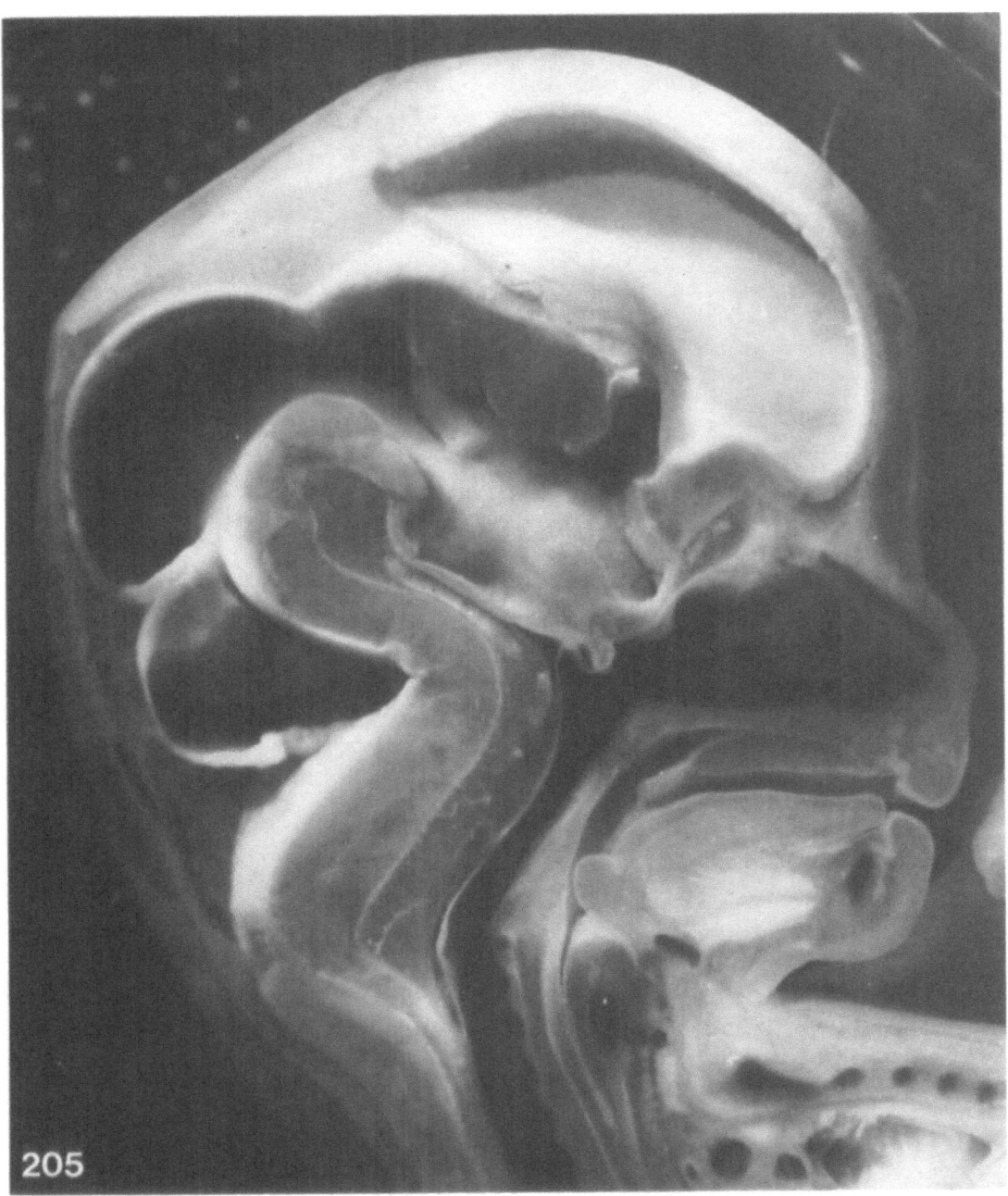

Figure 205. Sagittally dissected fetal head and neck. Stage 8-2. Structures depicted: nasal septum, cartilaginous base of skull, hypophyseal fossa, cervical vertebrae, upper and lower lips, plate, oral cavity, tongue, epiglottis, pharynx, laryngeal opening, and esophagus. Components of the CNS: Ventricle of cerebral hemisphere, corpus striatum, pallium, third ventricle with exposed hypothalamus, tuber cinereum, hypophysis, mamillary recess of third ventricle, broad mesencephalic aqueduct, cut fourth ventricle with superior medullary velum containing the rhombic lip (cerebellar primordium), choroid plexus of the fourth ventricle, pons, pontine flexure, and medulla oblongata.

110

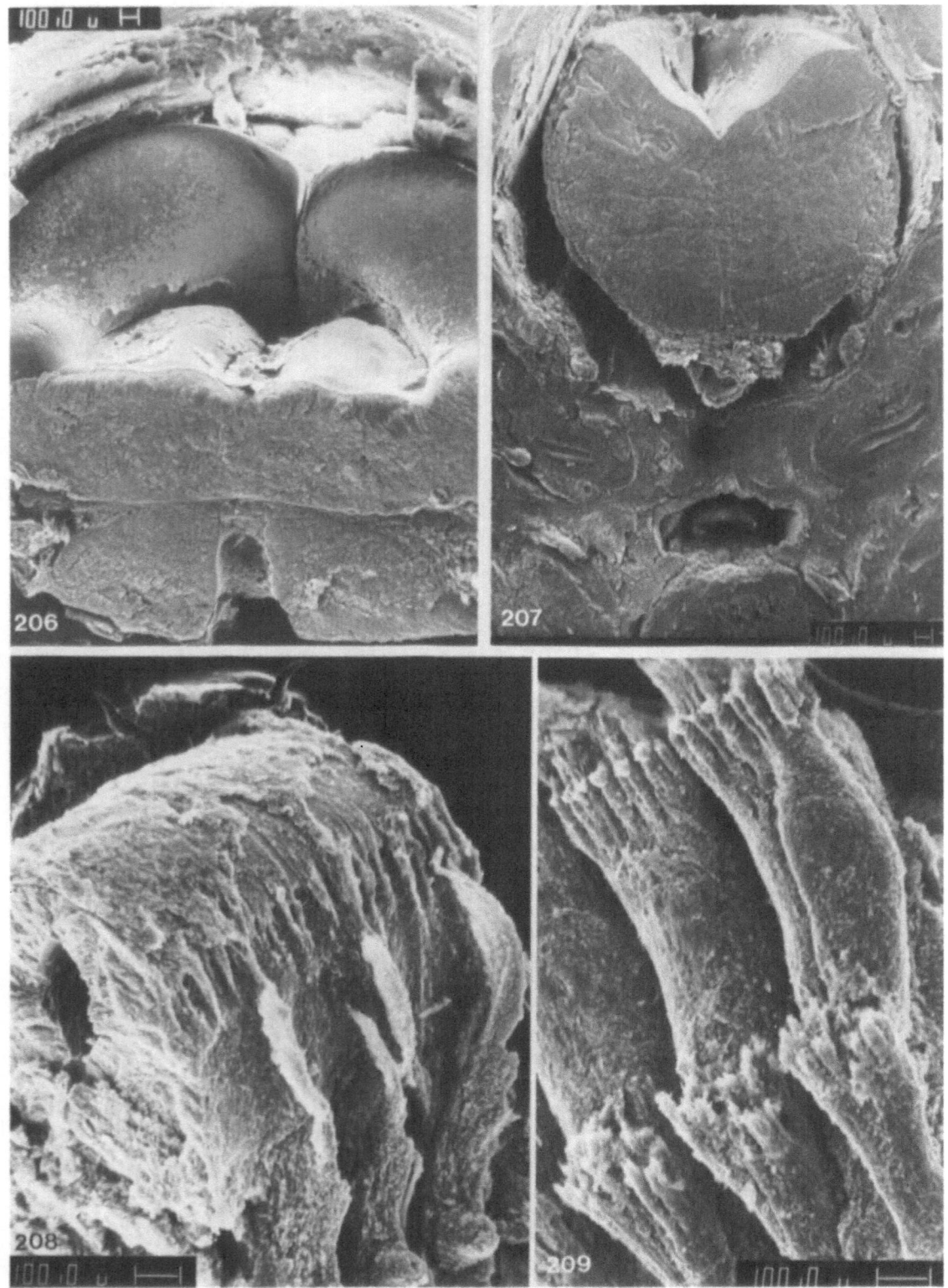

Figures 206–209. **(206)** Transversally dissected anterior portion of fourth ventricle; fusing rhombic lips. Stage 8-2. **(207)** Transversely dissected posterior portion of fourth ventricle; medulla oblongata. Stage 8-2. **(208)** Dissected medullary tube together with vertebral primordia. Dorsal extensions from vertebral primordia are mesenchymal condensations of neural vertebral processus. **(209)** Dissected thoracic spinal ganglia viewed from medullary tube. There are ca. 8–10 bundles of neural fibers constituting dorsal and ventral roots of the spinal cord.

the floor of the fourth ventricle. The metencephalon extends from the rhombencephalic isthmus to the pontine flexure. The myelencephalon begins at the pontine flexure and extends to the first spinal nerve. The roof plate of the fourth ventricle comprises the metencephalic portion (the anterior medullary velum, the rhombic lips, and the anterior portion of the posterior medullary velum) and the myelencephalic portion (the posterior portion of the posterior medullary velum). The rhombic lips fuse dorsally over the fourth ventricle into the cerebellar plate. The medical portion of the cerebellar plate represents the vermis and the lateral portions are the cerebellar hemispheres. Consequently a deep transverse furrow separates a phylogenetically old cerebellar portion – the nodulus and left and right floculi. The vascular mesenchyme covered by ependymal cells protrudes into the cavity of the fourth ventricle as the tela choroidea of the choroid plexus. The choroid plexus extends from the midline foramen of Magendi laterally to the foramena of Luschka. The foramina (holes) are formed secondarily in the ependymal epithelium of the velum.

The spinal cord (Figs. 208–216)
On the developing medullary tube, three areas are recognized: the ventral plate (floor plate); the basal plate, containing motor neurons whose axons leave the spinal cord as the anterior (motor) roots; and the alar plate, containing sensory neurons receiving axons from the neurons located within the spinal ganglia outside the spinal cord. These axons enter the spinal cord as dorsal (sensory) spinal roots. As the medullary tube closes, the alar plates of both sides meet in the midline and fuse into a glial posterior median septum of the medullary cord. The medullary tube originally fills the entire medullary canal. Spinal nerves leave and enter the tube in the level of each segment. During the fetal period, the vertebral column grows faster than the medullary cord and therefore the medulla ascends. Its caudal end finally reaches the level of L_2. The nerve fibers located within the medullary canal below the level of L_2 are known as the cauda equina.

The peripheral nerves (Figs. 63c, 64c, and 209)
Peripheral nerves are bundles of neural fibers (afferent and efferent processus of ganglionic cells).

The mixed spinal nerves divide into the dorsal and ventral primary rami: the dorsal ramus grows into the dorsal portion of the adjacent myotome and branches under the dorsal body wall; the ventral ramus supplies the ventral portion of the adjacent myotome and the ventral body wall. The development of the myotomes into muscles and their innervation are closely related.

The neural crest (Figs. 214 and 215)
As the neural tube closes, neuroectodermal cells migrate on the surface of the neural tube into the space underneath the surface ectoderm. The sheath of neuroectodermal cells located dorsally from the neural tube after its closure is known as the neural crest, which extends from the mesencephalon to the sacrococcygeal region. Most neural crest cells develop into spinal ganglia, while others constitute the autonomic nervous system, argentaffin cells, Schwann cells, pigment cells, odontoblasts, and most of the head and neck mesenchyme.

The sense organs (the eyes and the ears)

The eyes (Figs. 87c–89c and 217–226)
The eyes originate from two diencephalic neuroectodermal evaginations: the optic vesicles. The optic vesicles induce differentiation of adjacent surface ectoderm into the lens placode. The neuroectoderm gives rise to the retina, and the surface ectoderm to the lens, to the anterior epithelium of the cornea, and to the epithelium of the conjuctiva. All other components of the eye are mesenchymal in origin.

The optic vesicles represent lateral expansions of the diencephalic wall anterior to the rostral end of the notochord. The lateral wall of each optic vesicle contacts the surface ectoderm, inducing formation of the lens placode. Consequently the lateral wall of the optic vesicle invaginates and the optic vesicle changes into the optic cup. Coincidently the optic placode transforms into the lens groove, which closes into the lens vesicle and becomes detached from the surface ectoderm. The lens vesicle is located inside the optic cup behind its opening. The hyaloid artery comes to the eye on the underside of the optic cup, which is temporarily inferiorly incomplete. The deep notch on the underside of the

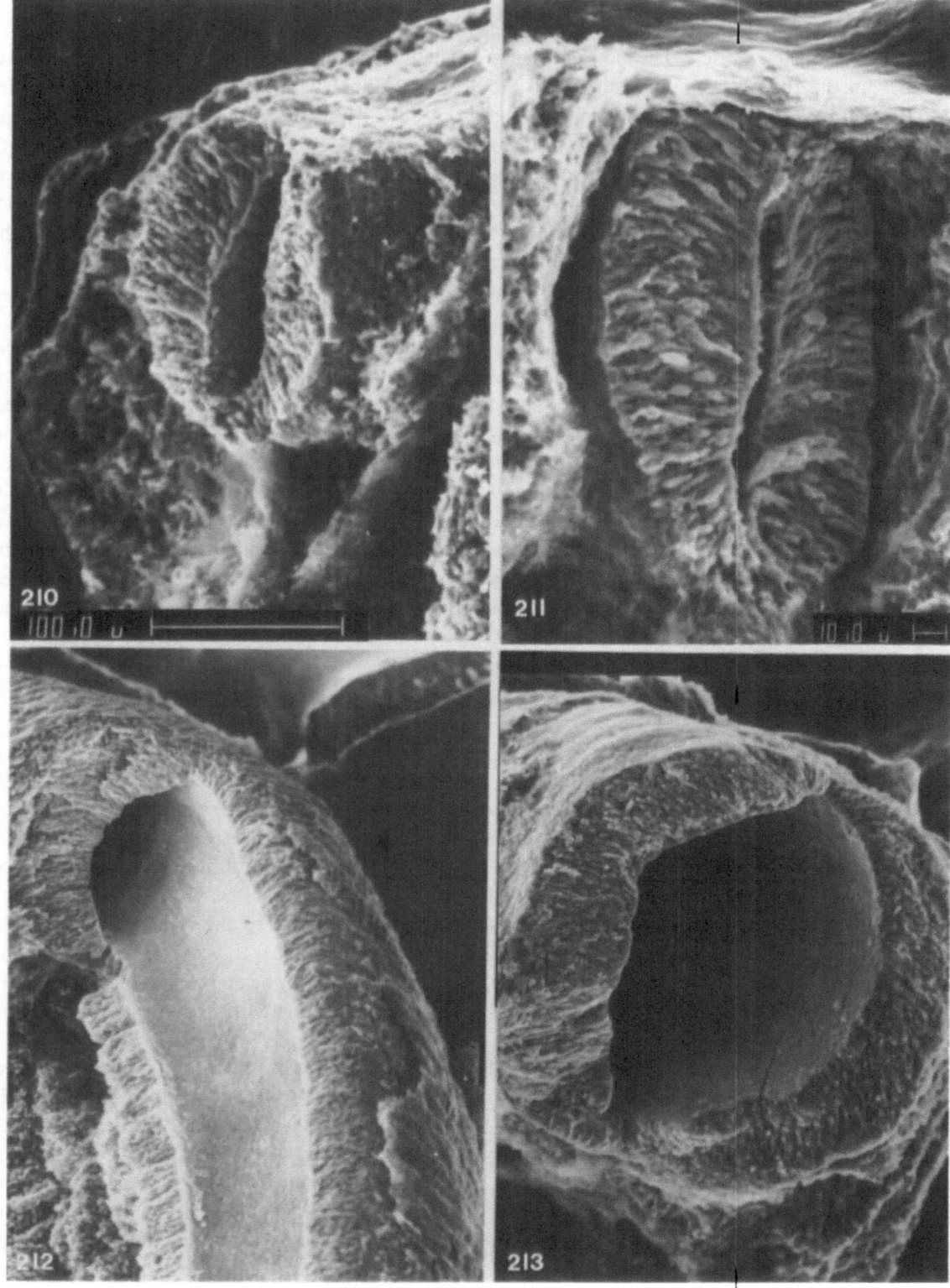

Figures 210–213. **(210)** Closed medullary tube. Stage 6-2. **(211)** Medullary tube ventrally contacting notochordal plate; dorsally, formation of neural crest. Stage 6-2. **(212)** Longitudinally dissected neural tube. Stage 7-2 **(213)** Dissected neural tube with a wide central canal. Stage 7-2.

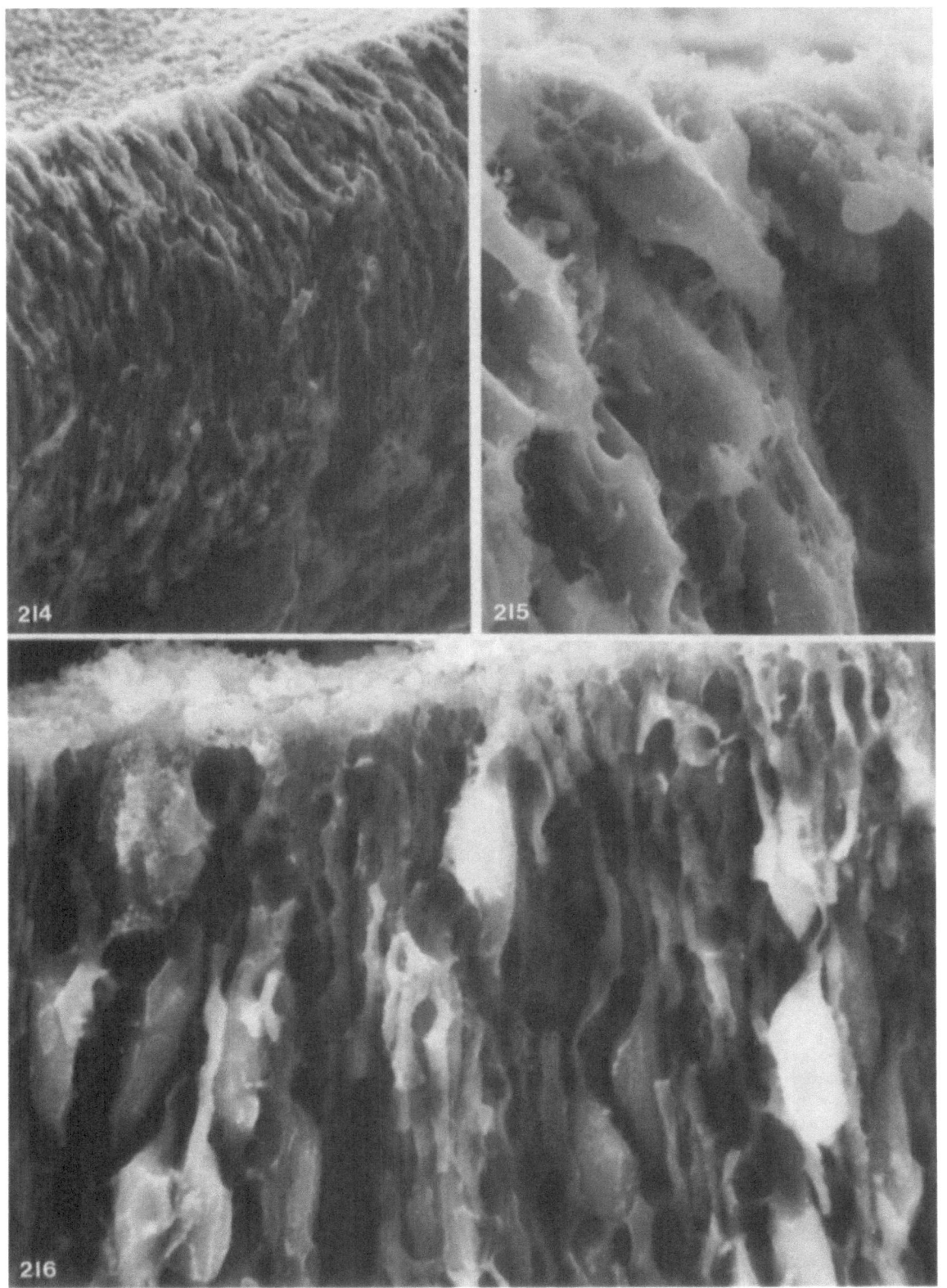

Figures 214–216. **(214)** Closed neural tube with neural crest cells on its dorsal surface. Stage 7-2. **(215)** Cells of the neural crest. Stage 7-2. **(216)** Neuroepithelial tissue of the medullary tube. Stage 7-4.

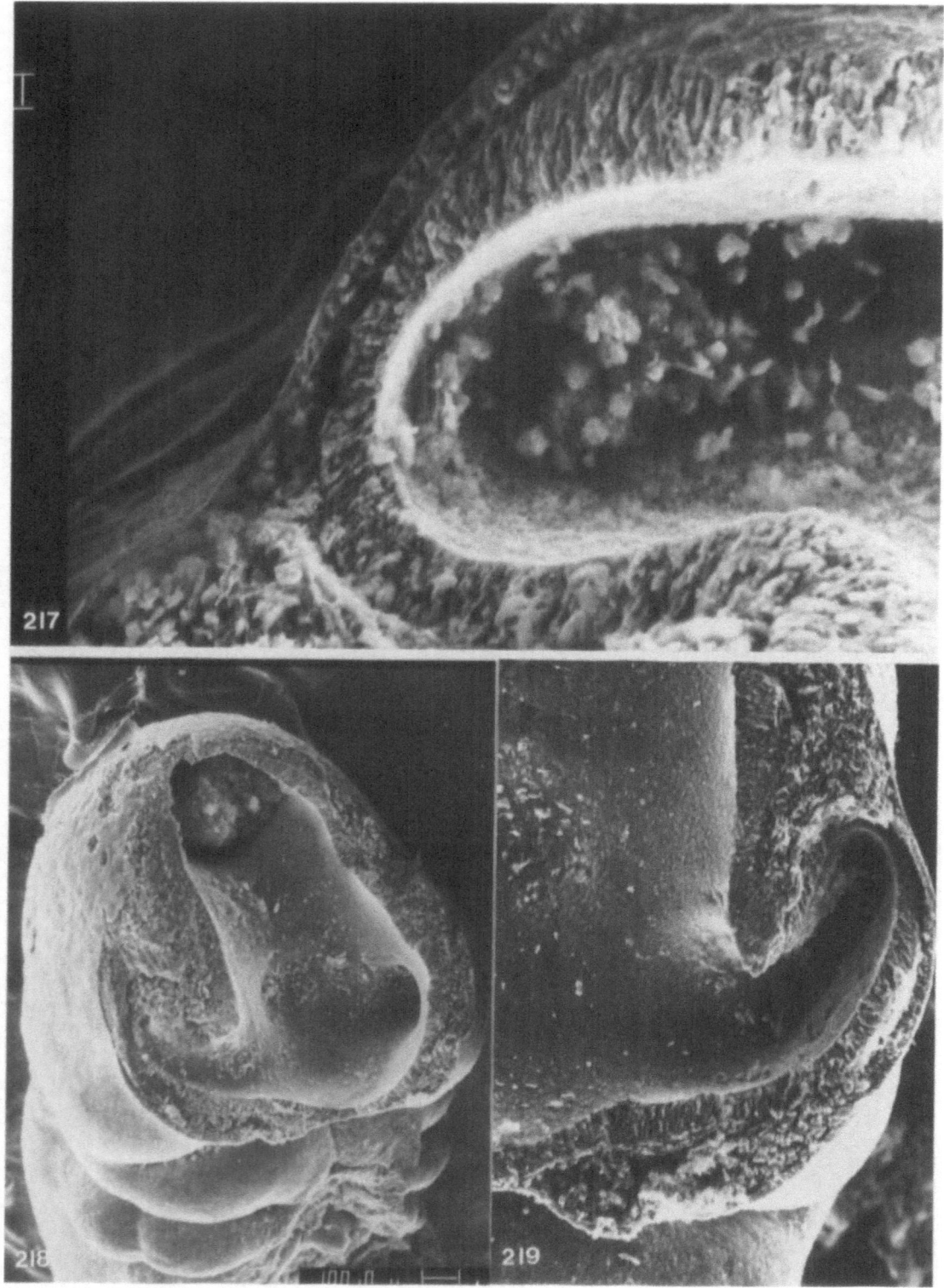

Figures 217–219. (217) Dissected optic vesicle contacting surface ectoderm. Stage 6-3 **(218)** Dissected prosencephalon with optic vesicles. Optic vesicles are connected with future diencephalon by broad optic stalks. Stage 7-2 **(219)** Dissected optic vesicle. Stage 7-2. Amorphous substance deposited between lateral wall of optic vesicle and surface ectoderm precedes formation of the lens vesicle.

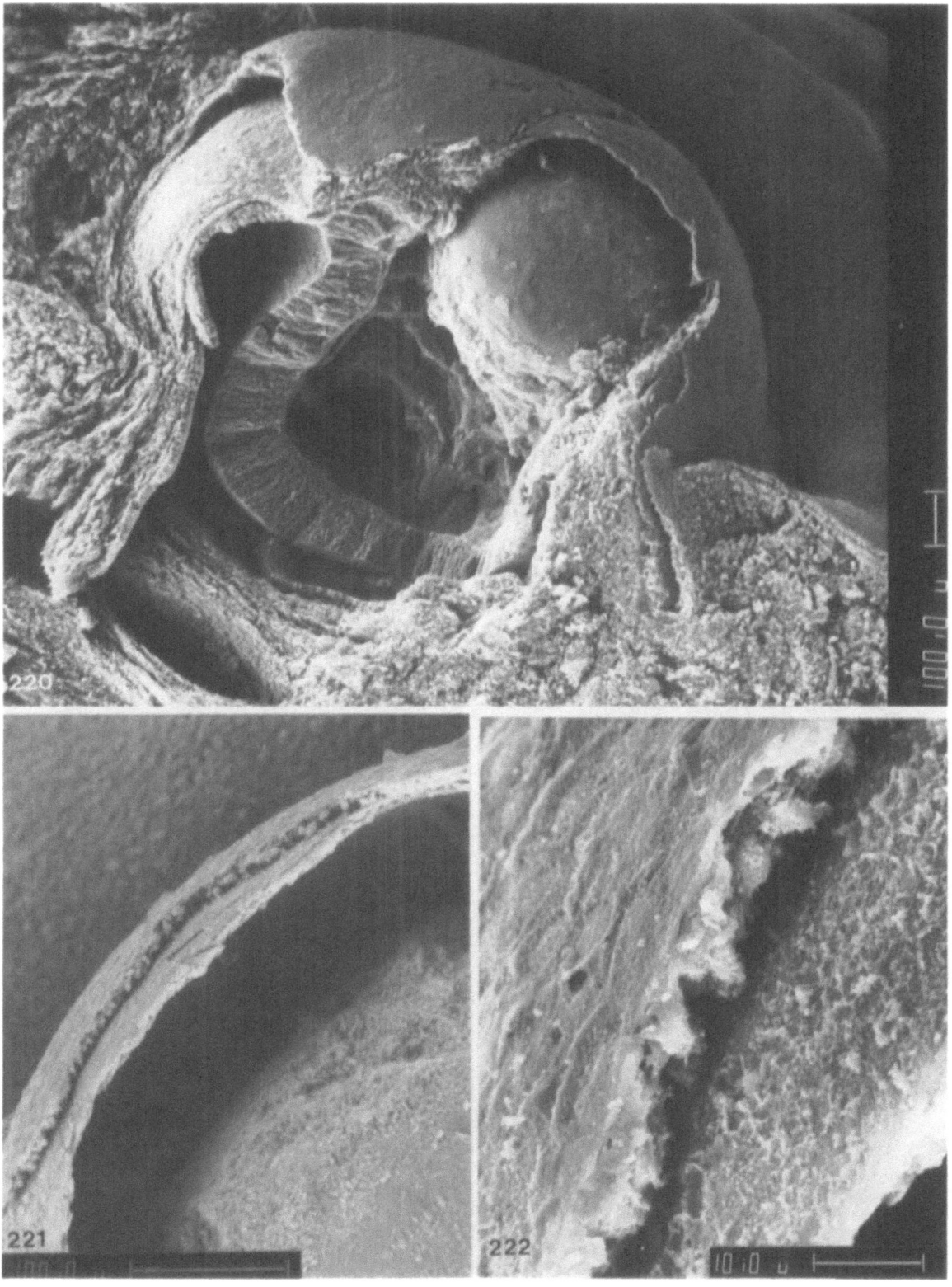

Figures 220–222. The eye. Stage 8-1 **(220)** Dissected eye in situ. Structures depicted: cornea (anterior epithelium and stroma), sclera, conjunctival sac, anterior eye chamber, lens, vitreous chamber filled with artificially precipitated vitreous humor, and artificially separated neural and pigment layers of the retina. **(221)** Cornea (surface epithelium, stroma): anterior chamber and surface of lens. **(222)** Anterior corneal epithelium and surface of corneal stroma.

116

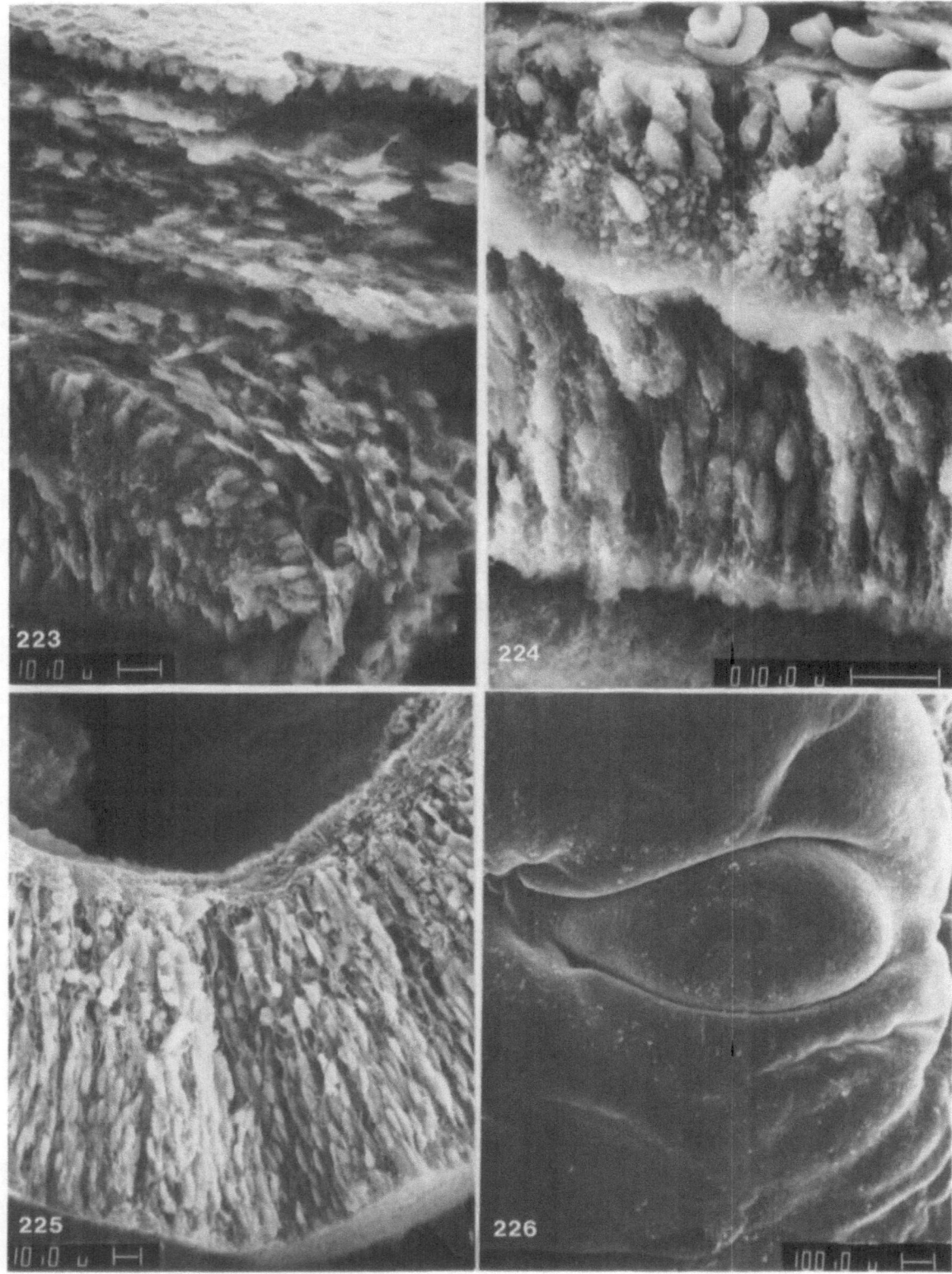

Figures 223–226. (223) Dissected eye. Cornea (epithelium and connective tissue of the propria), choroid (connective tissue), and margin of the optic cup with an inner and outer layer (ora serrata). **(224)** Dissected retina. Structures depicted: capillary with erythrocytes on the surface of retinal pigment layer, pigment layer of retina (numerous melanosomes within cytoplasm of cells), neural layer of retina, and precipitated vitreous. **(225)** Dissected neural layer of retina prior to differentiation of specific retinal elements. **(226)** Surface of the eye fissure. Stage 8-1.

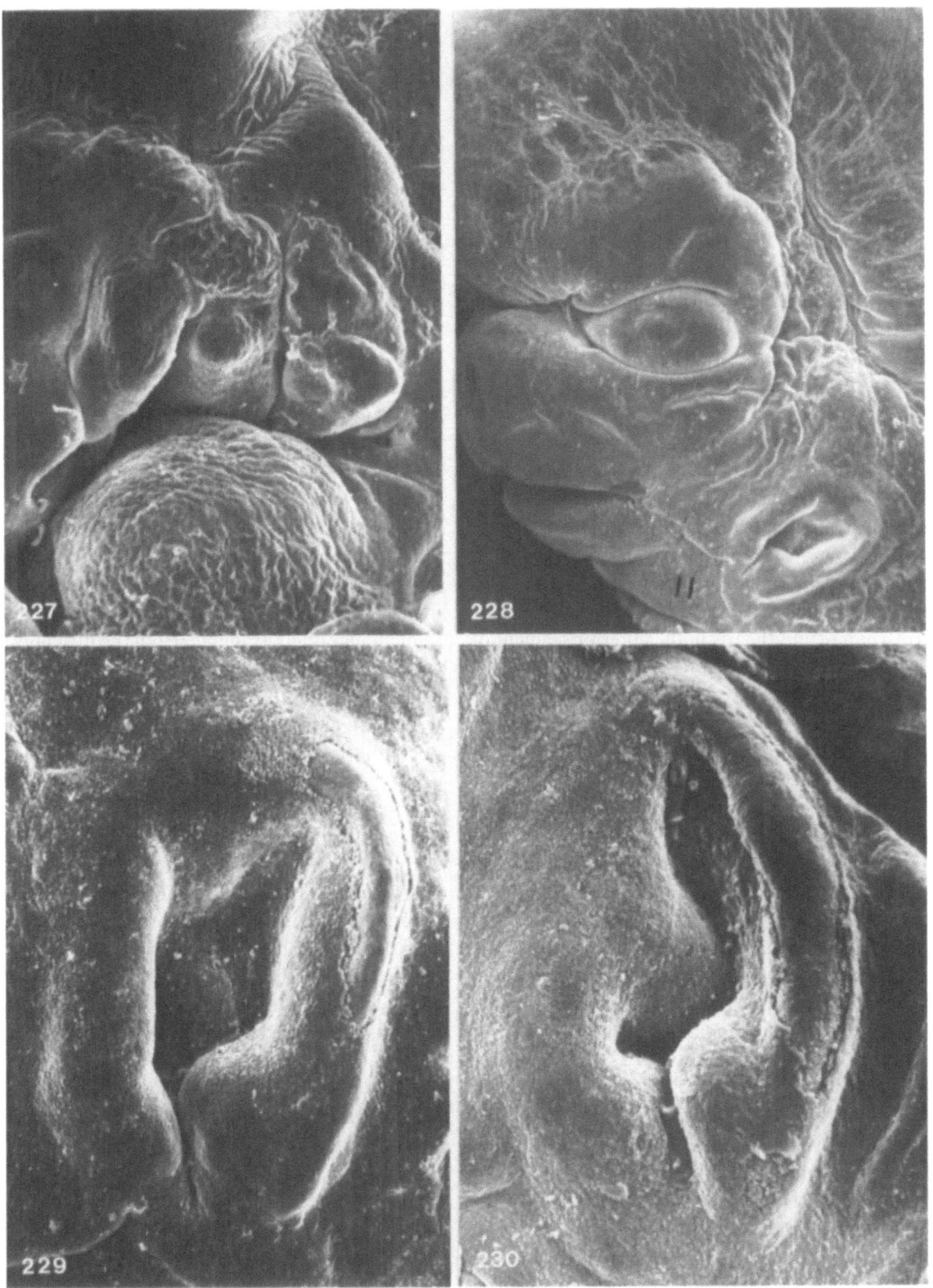

Figures 227–230. The external ear. **(227)** Mesenchymal tubercles of the first and second pharyngeal arches preceding development of the external ear. **(228)** Physiologically low-set ear (pinna). Stage 8-1 **(229)** The pinna, formed by mesenchyme from the first and second pharyngeal arches. Stage 8-1 **(230)** The pinna. Stage 8-2.

optic cup is known as the hyaloid fissure. As the optic cup expands, the margins of the hyaloid fissure fuse over the artery and the fissure disappears.

The neuroectodermal eye cup exhibits an outer and an inner epithelial layer. The outer layer gives rise exclusively to a single layer of pigmented epithelium. The inner layer in the posterior segment of the eyeball provides all remaining retinal layers of the optic portion of the retina. The blind portion of the retina, lining the inside of the anterior segment of the eyeball, remains double layered, does not develop any neuroreceptors, and becomes attached to the posterior surface of the corpus ciliare and to the iris.

The mesenchyme around the optic cup differentiates into an inner vascular layer and an outer fibrous layer. The vascular layer is formed only around the optic cup contributing to the uvea within the posterior segment of the eyeball, and the corpus ciliare and the iris in the anterior segment of the eyeball. The fibrous layer is formed around the whole eyeball and differentiates around the optic cup into the sclera and into the substantia propria of the cornea underneath the corneal surface epithelium. The space anterior to the iris and lens (between the fibrous and vascular layers of the eyeball) is the anterior eye chamber. The amorphous substance filling the optic cup behind the lens is the vitreous body. During its fetal period, the hyaloid artery runs through the vitreous body supplying the lens. The artery regresses before birth, leaving the lens avascular. The space between the vitreous body and the posterior surface of the iris is the posterior eye chamber.

The optic nerve's primordium is a double-layered tube – the stalk of the optic cup – with the hyaloid artery in the center. The axons from the retinal ganglionic cells grow along the epithelium of the optic stalk. The axons from the lateral portions of the retina run laterally and do not cross. The fibers from the medial retinal portions cross in the optic chiasma.

The eyelids are skin duplicatures supported by fibrous tarsal plates of mesenchymal origin. The margins of the eyelids fuse at days 56–60; the fused lids are temporarily separated by an epithelial plate, which disintegrates at the end of 26th conceptional week.

The ears (Fig. 90c)
The internal ear originates from a special neuroectodermal otic placode. The middle ear and the external ear are derivatives of the first and second branchial arches and of the first branchial pouch.

The internal ear. Two neuroectodermal otic placodes appear in the surface ectoderm laterally to the open rhomboencephalic vesicle. The placodes deepen into grooves, close, and subsequently detach from the surface ectoderm as the otic vesicles (otocysts). At the same period, neuroblasts detach from the closing rhombencephalon (rhombencephalic neural crest) and from statoacoustic (vestibulocochlear) ganglion in the vicinity of each otocyst. Each otocyst divides into a saccular and a utricular portion. Three pocket-like primordia of semicircular ducts evaginate from the utriculus. The central portions of semicircular evaginations come together and disappear. The peripheral portions of each evagination transform into the semicircular duct: the lateral semicircular duct has its proper opening into the utriculus; the superior and posterior semicircular ducts share a common opening.

From the saccular portion grows a spiral of the cochlear duct comprising at seven months prenatally two and half of twist. The cochlea is formed originally by a single-layered cylindrical epithelium. The basal wall gives rise to the Corti organ. The connection of the utriculus and sacculus, from which an endolymphatic sac evaginates, becomes a narrow utriculosaccular duct.

The middle ear. The cavity of the middle ear, and the pharyngeal tube (of Eustachius), are derivatives of the endodermal first pharyngeal pouch. The middle-ear ossicles develop within the mesenchyme and protrude secondarily into the middle-ear cavity. The incus and the malleus are derived from the first pharyngeal arch, the crura of the stapes from the second arch, the stapidial basis from the otic capsule.

The external ear (Figs. 227–230). The external ear comprises the pinna and the external auditory meatus. The pinna originates from approximately six ectoderm-covered mesenchymal tubercles of the first and second branchial arches. The three tubercles of the first arch give rise to the tuberculum tragicum, the tuberculum anterior hellicis and the tuberculum intermedium hellicis. The cauda hellicis, the tuberculum antitragicum, and the tuberculum lobulare are derivatives of the second pharyngeal arch.

The external auditory meatus represents the first ectodermal branchial pouch. The ectodermal obturatory membrane of the pouch changes into an epithelial plate, which disintegrates in the second half of pregnancy, leaving the terminal portion of the external auditory meatus and the portion of the external eardrum epithelium that is attached to the malleus. The eardrum is derived mostly from the obturatory membrane of the first branchial pouch.

The face, nose, and palate (Figs. 91c–96c and 231–263)

The face develops from structures located around the stomodeum: the frontofacial eminence and the first pharyngeal arches. The primitive stomodeum is a deep fold between the frontofacial eminence and the first pharyngeal arches with the pharyngeal membrane on its bottom. The frontofacial eminence is a bulging ectoderm-covered area over the closing forebrain. The first pharyngeal arches delineate the stomodeum ventrally and laterally, and differentiate into maxillary centers and mandibular arches. Mandibular arches of both sides fuse into a single primitive mandibular primordium. After development of maxillary centers, the stomodeum becomes pentagonal, delineated ventrally by mandibular arches, laterally by maxillary primordia, and dorsally by the frontofacial eminence. The pharyngeal membrane disintegrates and disappears, leaving the opening into the foregut. On the frontofacial eminence, the two optic areas with the lens placodes (induced by optic vesicles) appear laterally, and the two olfactory placodes (closely related to the development of cerebral hemispheres) are located anteriorly. Consequently, in the optic area, the lens placodes change into the lens vesicles and detach from the surface ectoderm. The eye fissures become delineated by the eyelids.

In the nasal area, mesenchymal proliferation around each nasal placode changes the nasal placode into the nasal groove delineated by a horseshoe-shaped nasal ridge. On each nasal ridge, a lateral portion (contacting the maxillary primordium), a medial portion, and a premaxillary portion can be distinguished. A deep linear depression between the lateral portion of the nasal ridge and the maxillary primordium is known as the nasolacrimal groove. On the bottom of each nasal groove, there is an olfactory dimple closed by the oronasal membrane (of Hochstätter). As the premaxillary portions of the nasal ridges bulge anteriorly into the primitive oral cavity and become located between the maxillary primordia, a deep linear depression, known as the nasal fin, appears anterior to each olfactory groove. Each nasal finn consists of a lateronasal–premaxillary portion and a maxillary–premaxillary portion. The lateronasal–premaxillary junction contains mesenchyme located anterior to the olfactory groove; the maxillary–premaxillary junction is temporarily represented by an epithelial plate, which is secondarily penetrated by mesenchyme as the maxillary primordium and the premaxillary portion of the nasal ridge fuse. A midline depression of the primary palate located between the premaxillary portions of the nasal ridges becomes filled by prechordal mesenchyme growing from an additional midline nasodorsal center. The nasodorsal center consists of an unpaired nasoseptal portion and a paired nasozygomatic portion. Cartilages of the so-called nasal capsule are formed within the mesenchyme of the nasodorsal center. The mesenchyme of the nasozygomatic portion of the nasodorsal centers fills the nasolacrimal grooves. The nasoseptal portion contributes to the nasal septum and the midline portion of the upper lip. According to the concept presented here, the dorsum and septum of the nose originate from the nasodorsal center, the alae of the nose from the lateral portions of the nasal ridges, and the colummela of the nose, delineating the nares medially, from the medial portions of the nasal ridges.

The primary palate is formed by maxillary primordia that fuse with the premaxillary portions of the nasal ridges. The prechordal nasoseptal mesenchyme, which 'glues' both premaxillary centers together, contributes (in the area of the primary

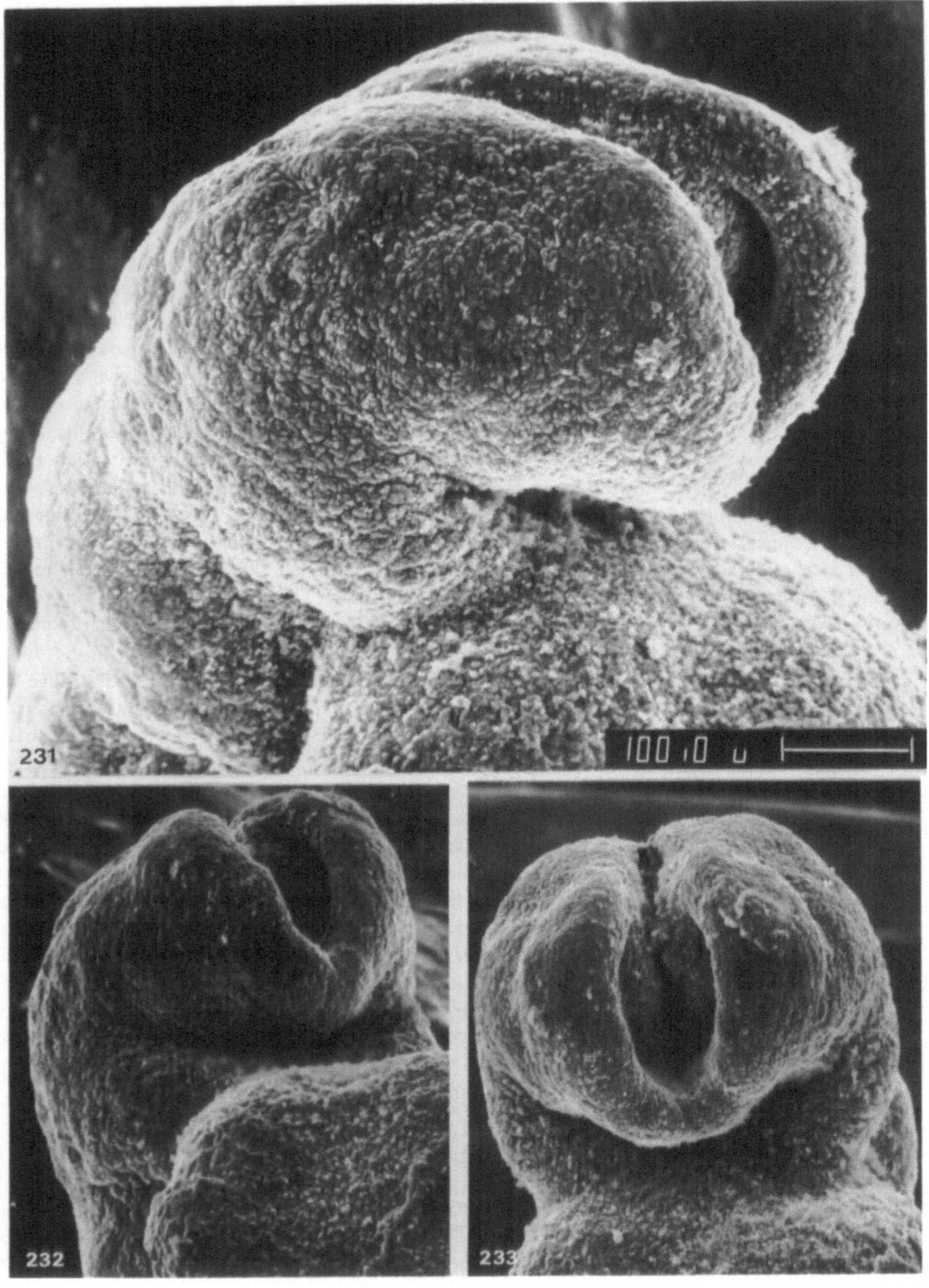

Figures 231–233. Embryonic head. Stage 6-2. Structures depicted: closing anterior neuropore, frontofacial prominence, first pharyngeal arches, oral fold closed by oral (buccopharyngeal) membrane, and heart bulge. **(231)** Right laterosuperior view. **(232)** Right lateroinferior view. **(233)** Anterior view.

123

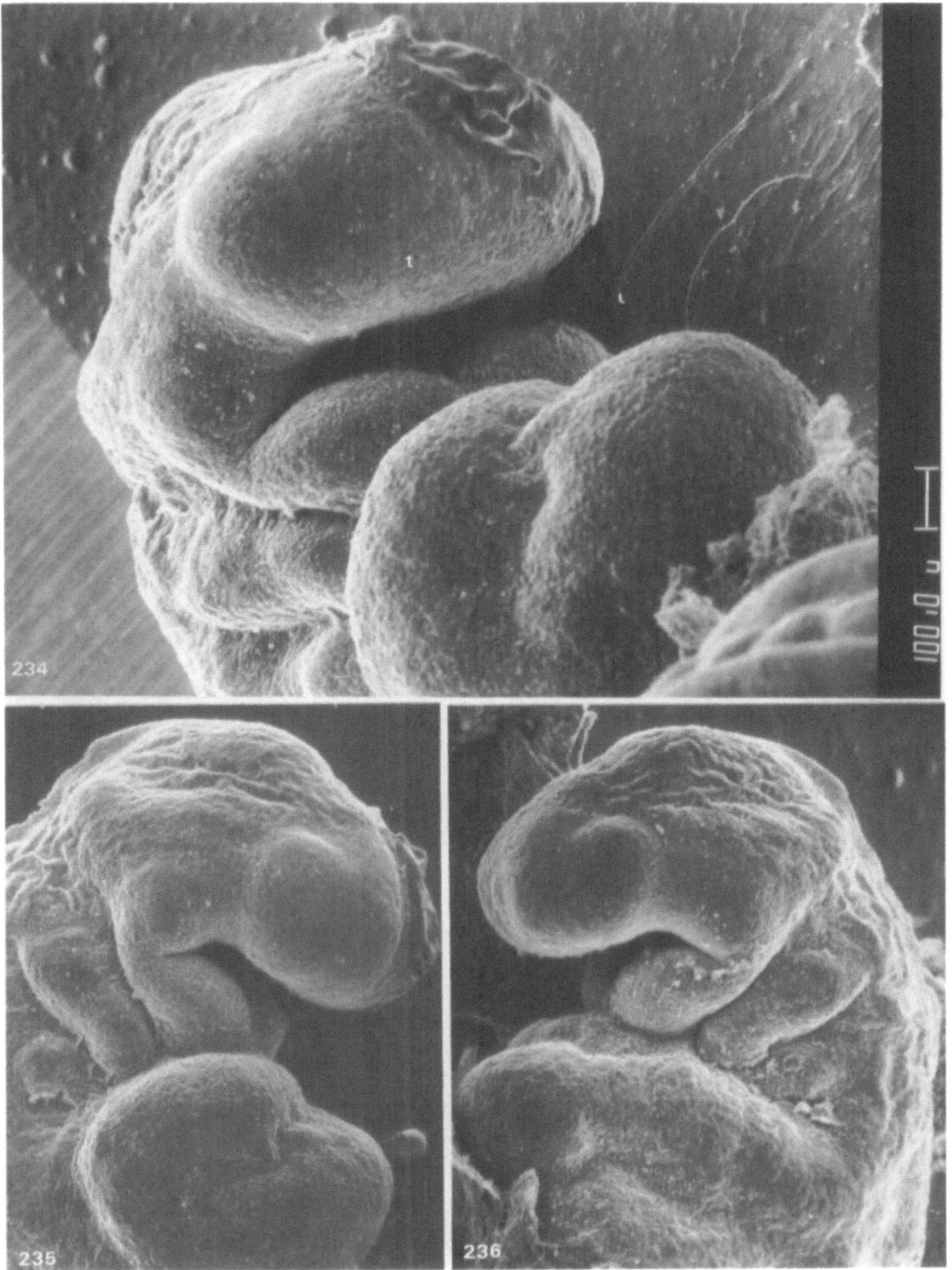

Figures 234–236. Embryonic head and neck. Stage 6-3. **(234)** Structures depicted: pentagonal stomodeum delineated superiorly by a frontofacial prominence with laterally bulging optic areas, laterally by maxillary portions of first pharyngeal arches, and anteriorly by mandibular portions of first pharyngeal arches. **(235)** Right lateral view. Structures depicted: frontofacial prominence with optic areas; maxillary and mandibular portions of first pharyngeal arch; stomodeum; second and third pharyngeal arches; first, second, and third ectodermal pharyngeal pouches; and heart bulge. **(236)** Left lateral view.

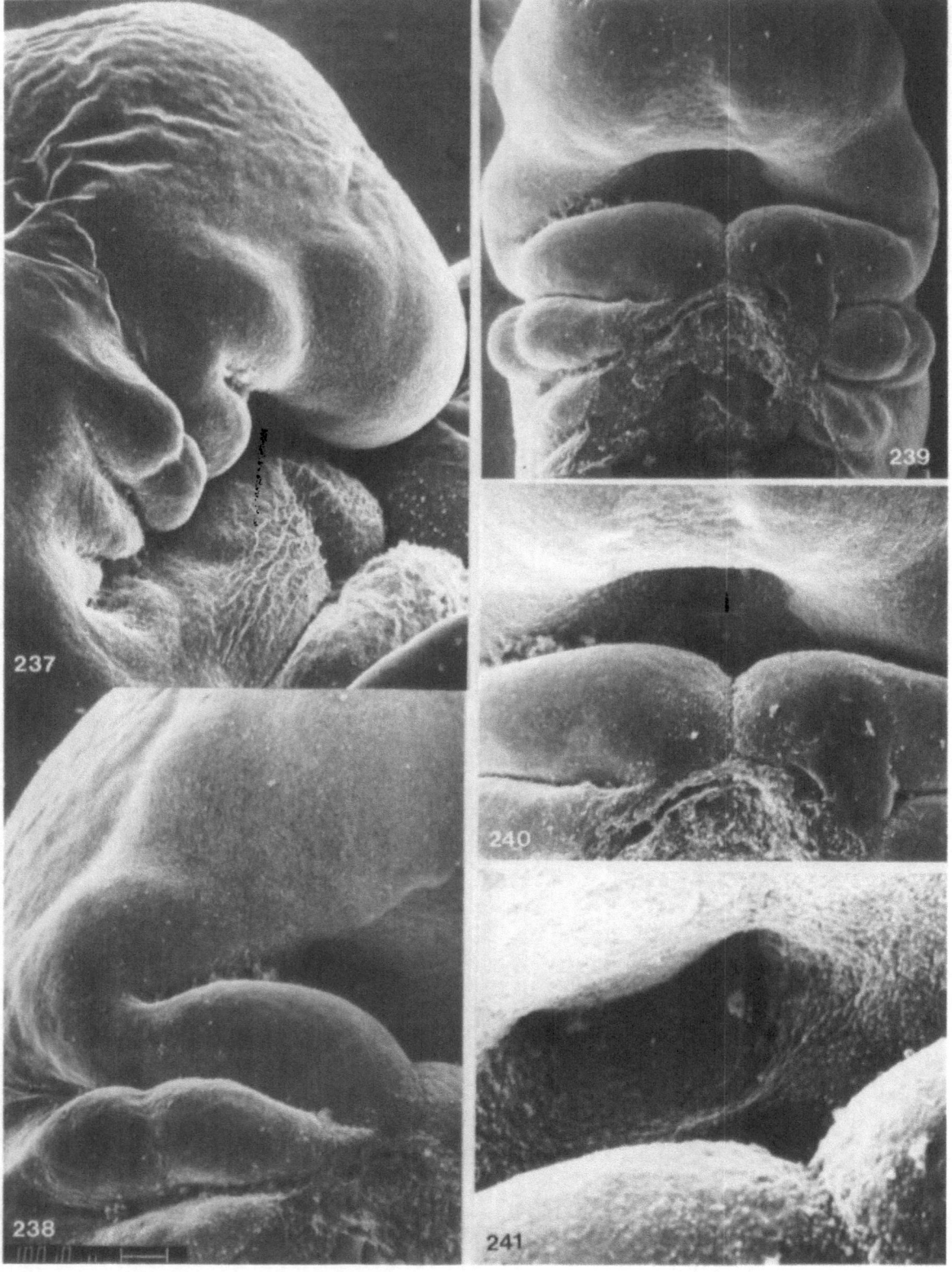

Figures 237–247. Orofacial area. Stage 7-1. **(237)** Right lateral view of the head. Structures depicted: frontofacial prominence with a bulging optic area; maxillary and mandibular portions of the right first pharyngeal arch; second pharyngeal arch with a distinct ventral and dorsal portions; and third, fourth, and fifth pharyngeal arches with corresponding ectodermal pharyngeal pouches. **(238)** Primitive oral cavity with a distinct opening of the hypophyseal (Rathke's) pouch, bulging optic area, first right pharyngeal arch, and second pharyngeal arch. **(239)** Anterior view of the head and neck. Primitive oral cavity is delineated by the frontofacial prominence and by maxillary and mandibular primordia of the first pharyngeal arches. **(240)** Fusing mandibular portions of the first pharyngeal arches. **(241)** Opening of the hypophyseal (Rathke's) pouch on the ceiling of the ectodermal portion of the oral cavity.

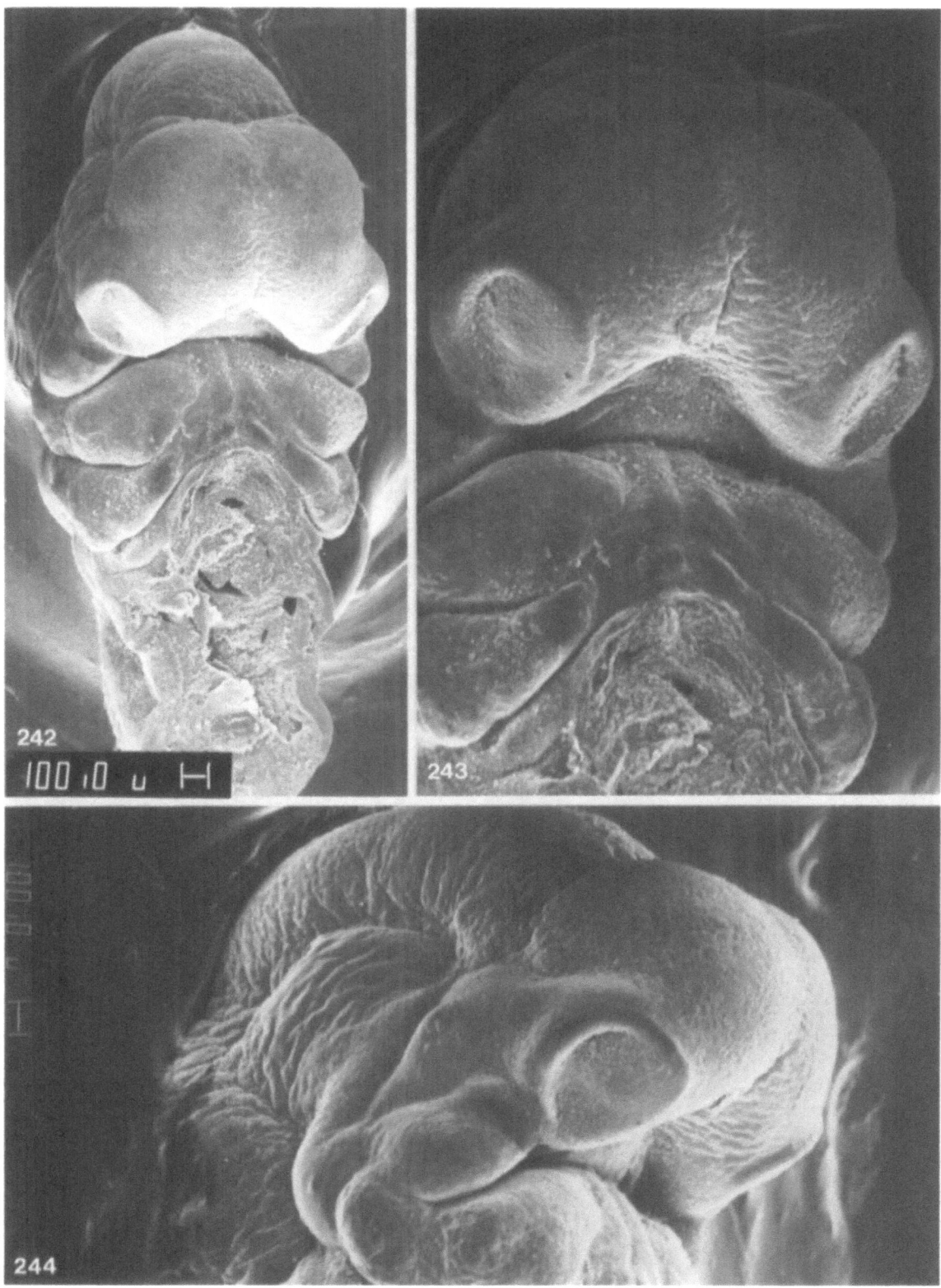

Figures 242–244. Embryonic head and neck. Stage 7-3. **(242)** Dissected head and neck. Stage 7-3. Anterior view. Structures depicted: frontofacial prominence with bulging cerebal hemispheres and olfactory placodes, maxillary primordia, oral cavity, mandibular arch, and second, third and fourth (rudimentary) pharyngeal arches. **(243)** Frontofacial prominence with olfactory placodes, and first, second, third, and fourth pharyngeal arches. Anteroinferior view. **(244)** Embryonic head. Lateral view. Structures depicted: frontofacial prominence with bulging cerebral hemispheres and olfactory placodes (early pits) delineated laterally by nasal ridges fusing with optic areas, maxillary and mandibular primordia, and primitive oral cavity.

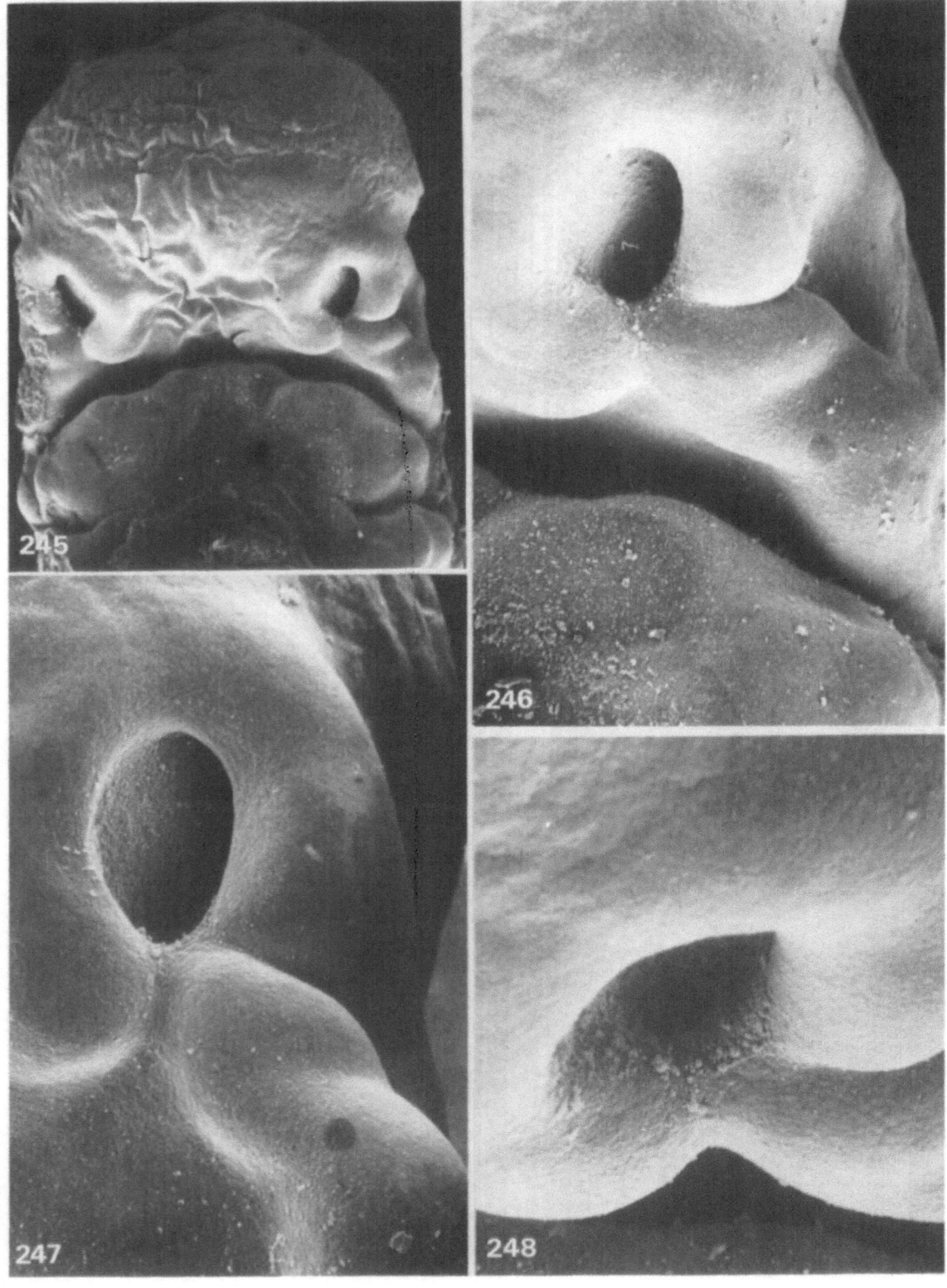

Figures 245–248. Nasal development. Stage 7-4. **(245)** Embryonic head with artificially shrunken surface of frontofacial prominence. Structures depicted: nasal pits delineated by nasal ridges, maxillary primordia, oral cavity, mandibular arch, and ectodermal first pharyngeal pouches. **(246)** Left nasal pit delineated by nasal ridge, maxillary primordium, nasolacrimal sulcus (laterally), and nasal fin (anteriorly, connecting nasal pit and oral cavity). Anterior view. **(247)** Nasal pit, nasal ridge, nasolacrimal sulcus, maxillary primordium, and nasal fin. Anteroinferior view. **(248)** Nasal pit with a distinct olfactory dimple (closed by an ectodermal nasopharyngeal membrane).

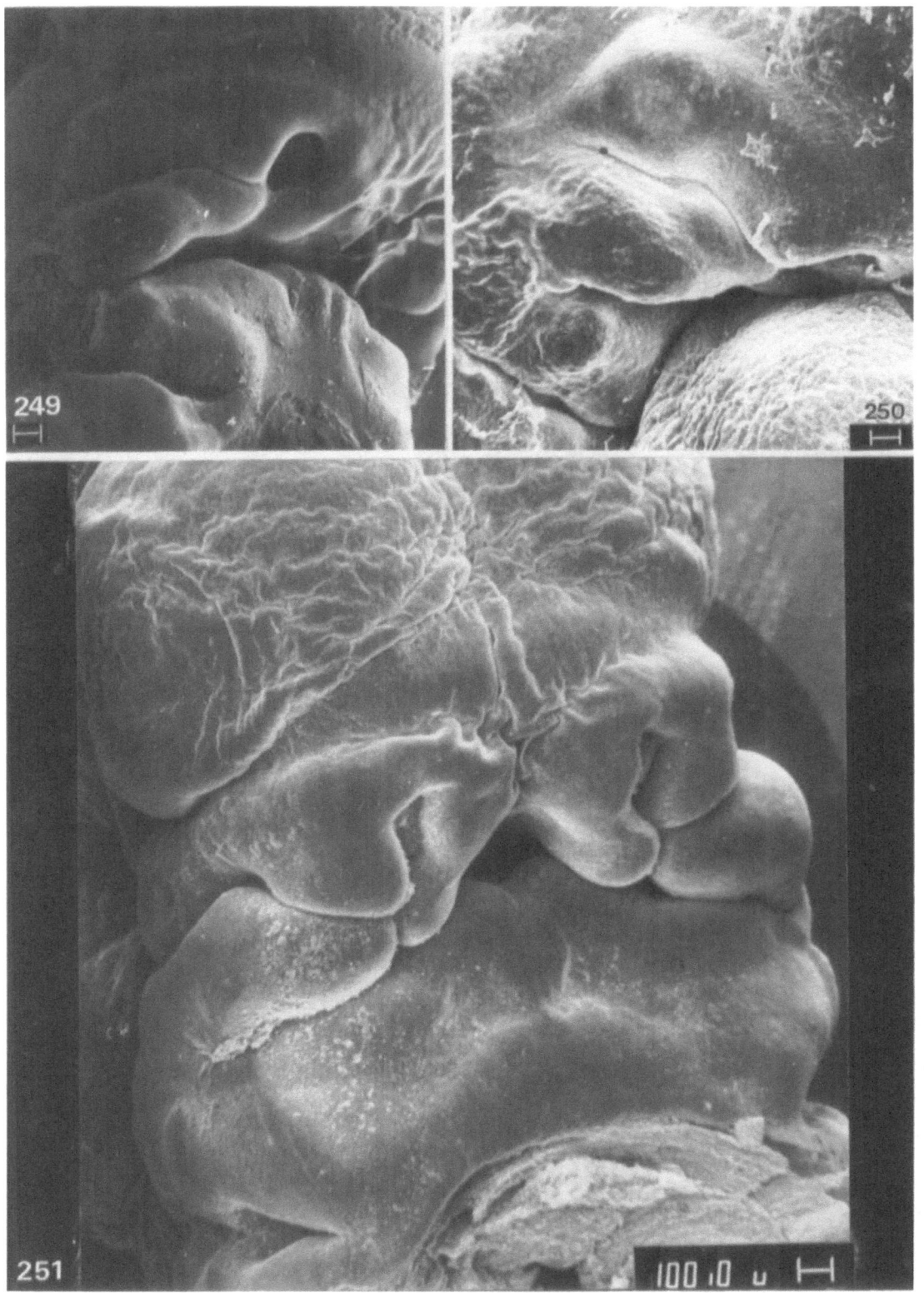

Figures 249–251. Embryonic frontofacial structures. Stages 7-4 and 7-5. **(249)** Opticonasal area: nasal pit delineated by nasal ridge, nasolacrimal sulcus, and maxillary and mandibular primordia. Stage 7-4. **(250)** Opticonasal area delineated inferiorly by nasolacrimal sulcus and lower eyelid, maxillary and mandibular primordium, and heart bulge. Stage 7-4. **(251)** Frontofacial area. Stage 7-5. Structures depicted: frontal area, nasodorsal center composed of a midline nasoseptal portion and laterally located nasozygomatic portions; opticonasal area: nasal pits delineated by nasal ridges with distinct premaxillary, medionasal, and lateronasal portions, nasolacrimal sulcus, maxillary primordium, and lower jaw.

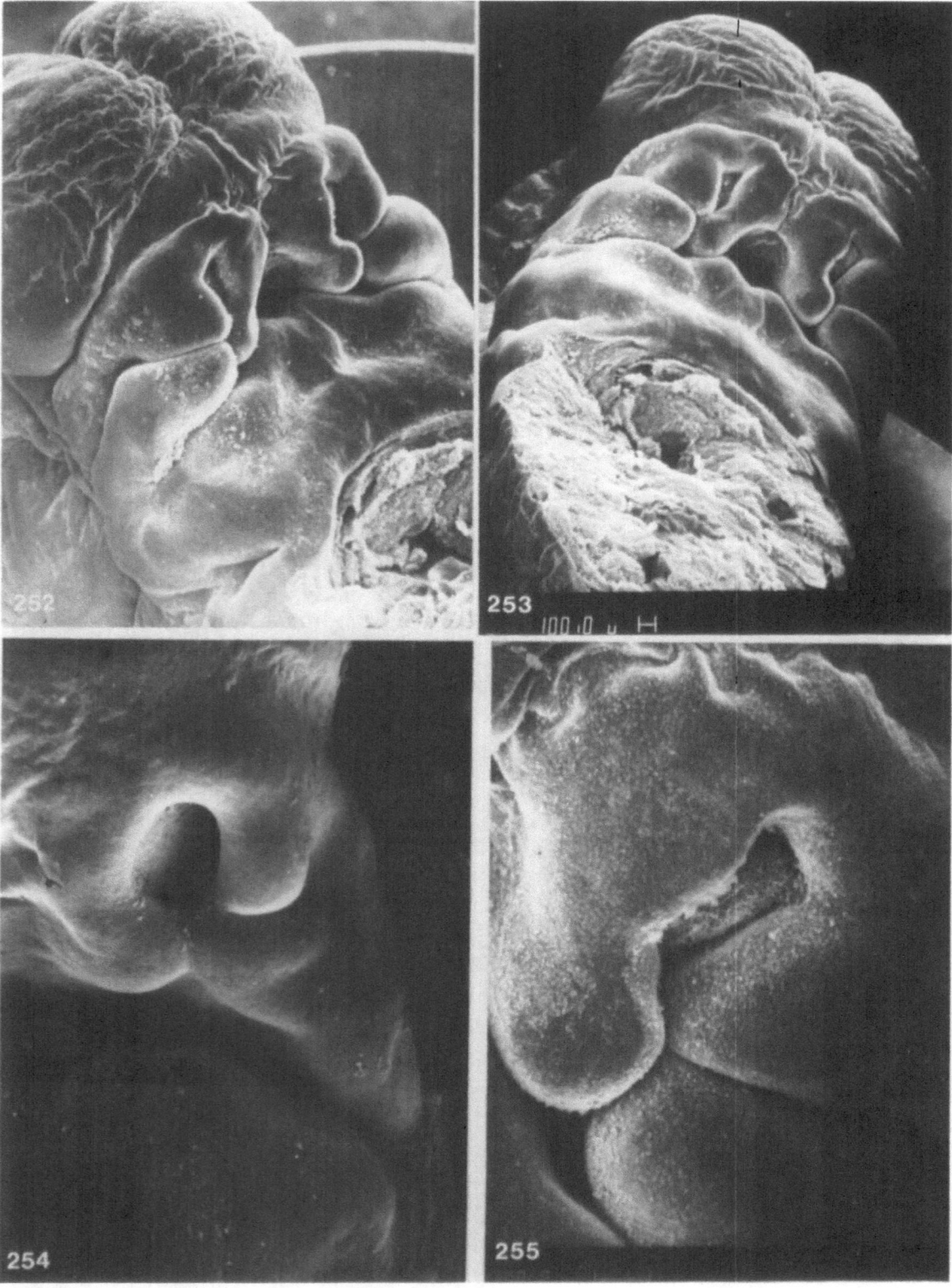

Figures 252–255. (252) Embryonic face. Stage 7-5. Right oblique view. Structures depicted: frontal prominence with bulging cerebral hemispheres, nasodorsal center (with an unpaired nasoseptal portion and paired nasozygomatic portion), opticonasal areas, nasal ridges, nasal pits, maxillary primoridum, lower jaw, and primitive external auditory meatus. **(253)** Embryonic face. Stage 7-5. Right inferior view. Structures depicted: frontal area with bulging cerebral hemispheres; nasodorsal center with nasoseptal and nasozygomatic portions; nasal pits delineated by nasal ridges with premaxillary, medionasal, and lateronasal portions; maxillary primordia; interpremaxillary depression; premaxillary–maxillary junction; and lower jaw. **(254)** Premaxillary–maxillary junction (nasal fin) at a primitive stage 7-4. **(255)** Premaxillary-maxillary junction at a progressed stage with lateronasal-premaxillary and maxillary-premaxillary portions. Stage 7-5.

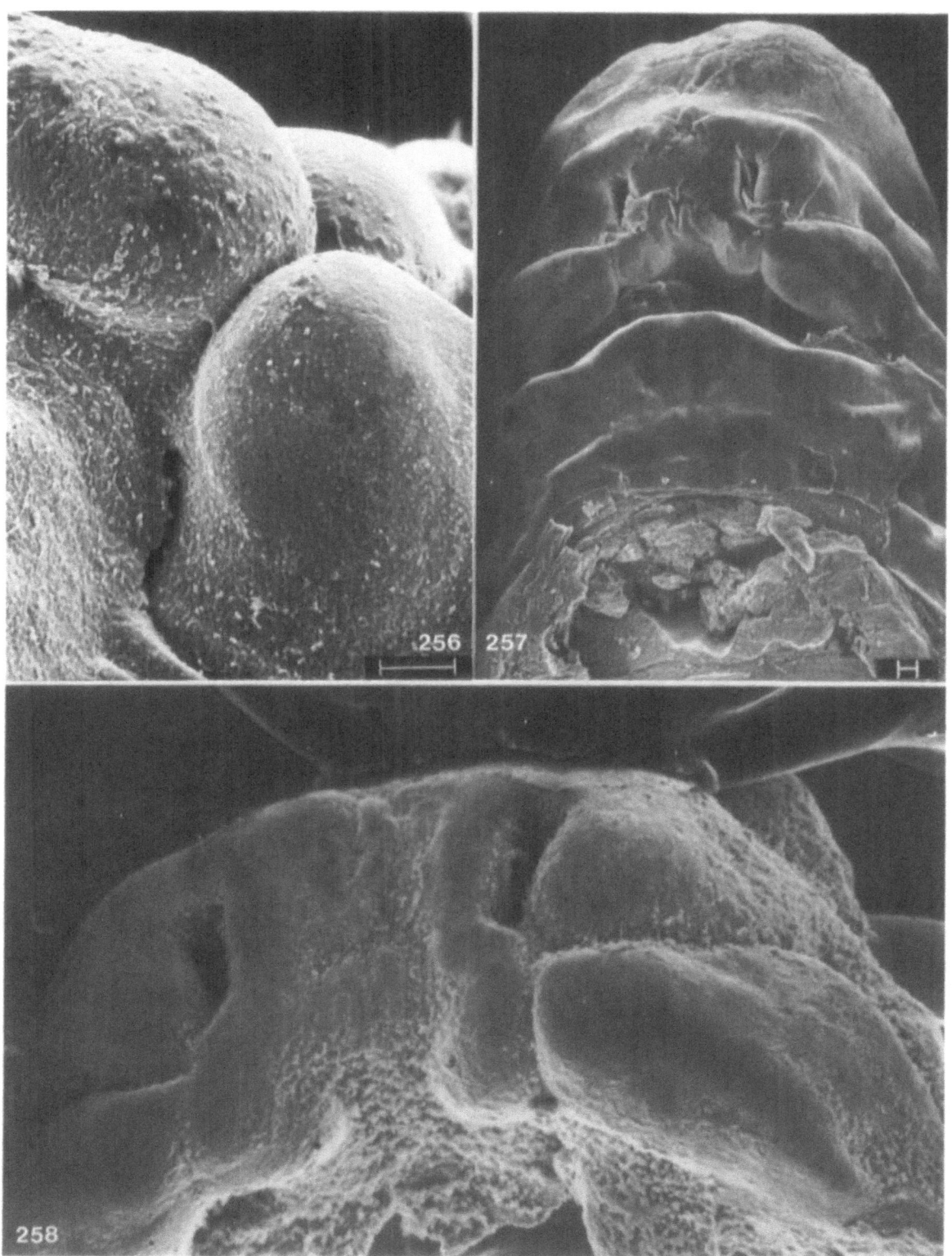

Figures 256–258. Formation of the primary palate. **(256)** Fusing premaxillary portion of nasal ridge with maxillary primordium; slit-like opening of a primitive choana. **(257)** Fusing premaxillary portions of nasal ridge with maxillary primordia (future upper lip), medial portions of nasal ridges (delineating medially external nares), and lateral portions of nasal ridges (delineating external nares laterally and contributing to nasal ala). Stage 7-5. **(258)** Upper lip. Stage 7-5. Anteroinferior view. Structures: depicted: premaxillary, medial, and lateral portions of nasal ridges; external nares; and maxillary primordia.

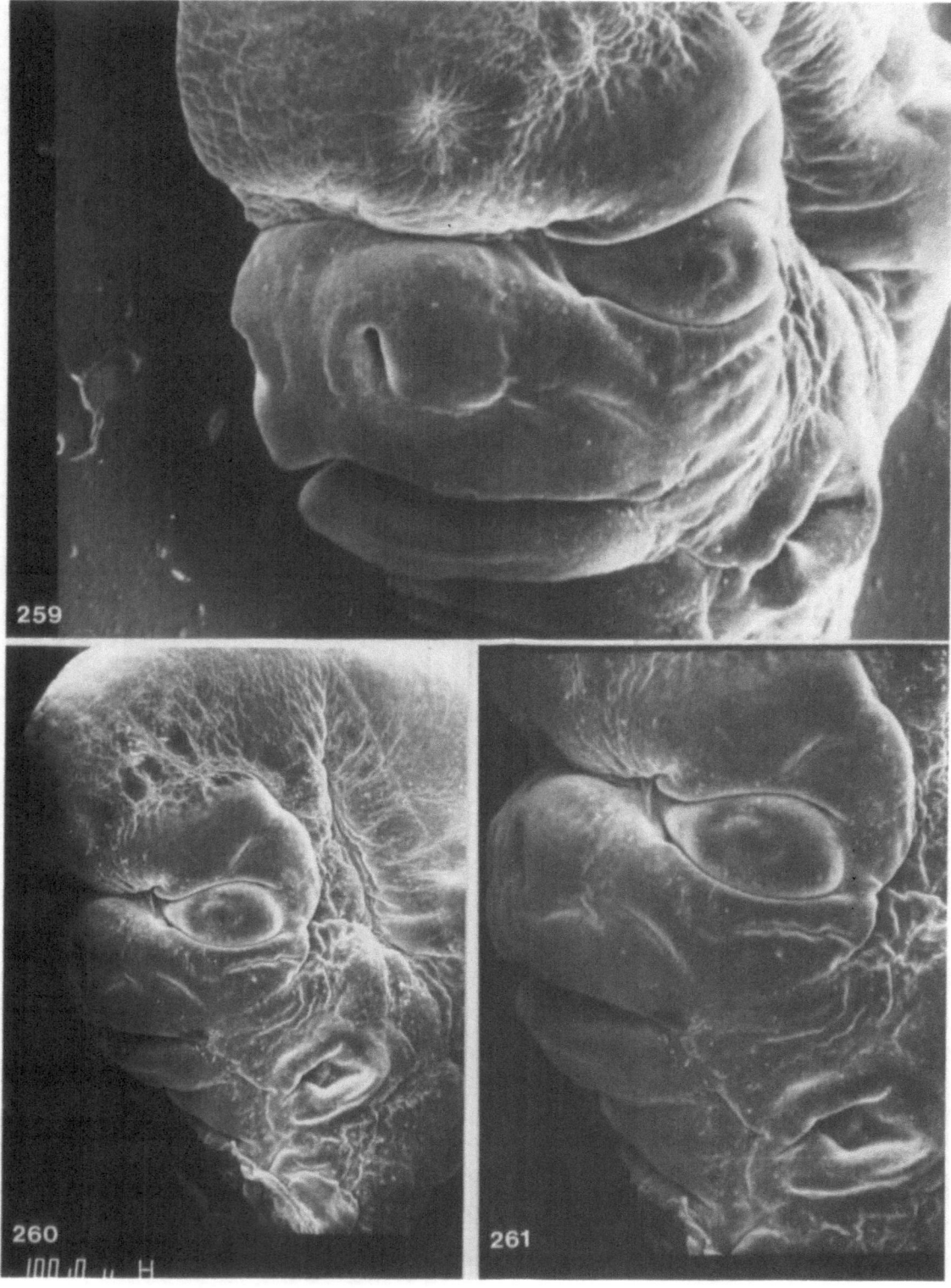

Figures 259–261. Embryonic face. Stage 8-1. **(259)** Oblique left anterior view. Structures depicted: frontal area, interocular depression (artificially deepened due to shrinkage of the specimen), eyes delineated by eyelids, cheeks (contributed to by nasozygomatic mesenchyme), nose, closed upper lip, lower jaw, and external ear (physiologically low set). **(260 and 261)** Left lateral views.

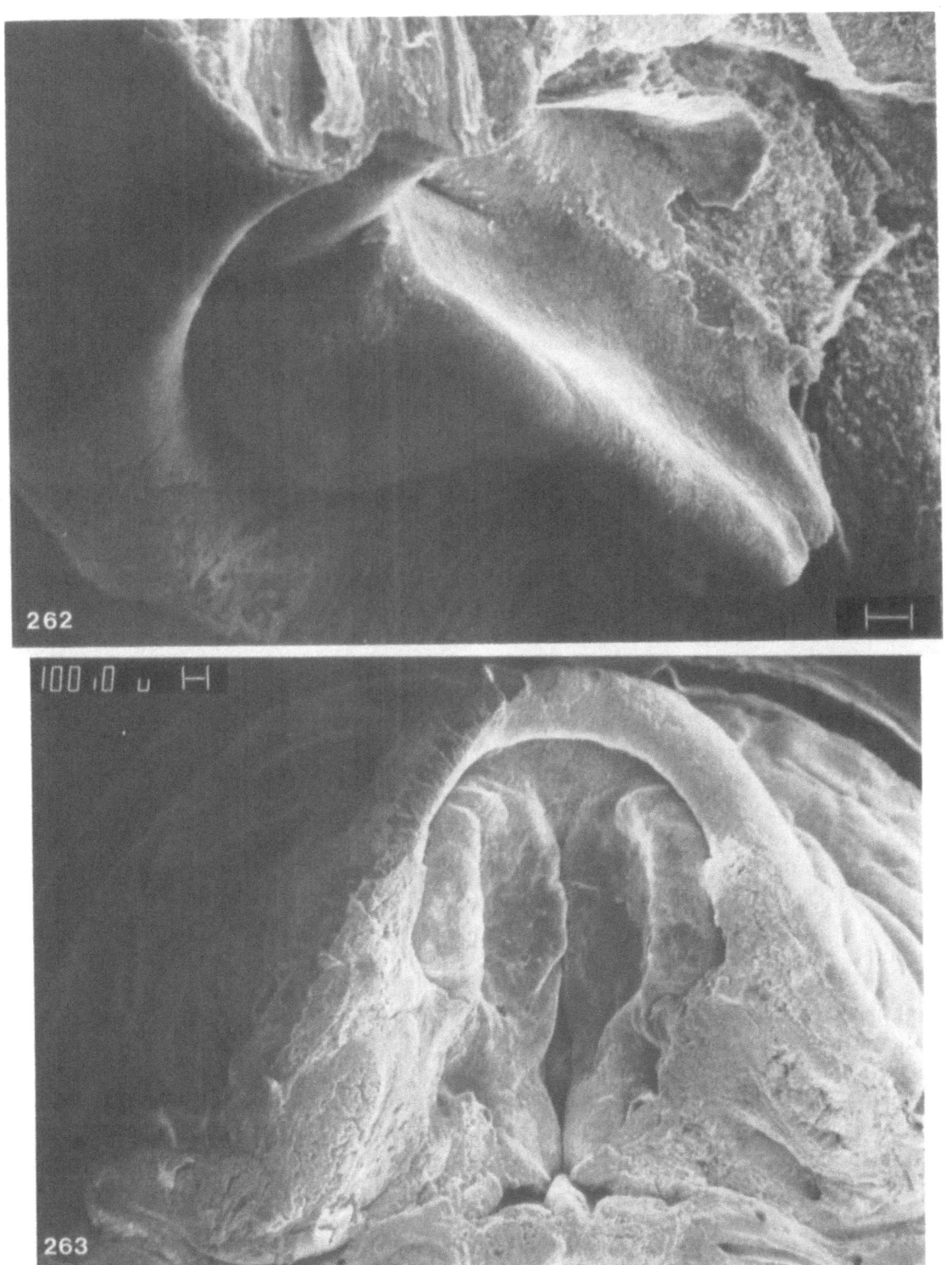

Figures 262 and 263. Dissected upper lip and secondary palate. **(262)** Sagittal dissection. Stage 8-1. Structures depicted: upper lip, premaxillary area of the palate, and sagittally oriented right palatinal shelf. **(263)** Dissected mouth cavity. Stage 8-2. Structures depicted: upper lip, premaxillary area of the palate, and merging palatinal shelves.

palate) to a small midline portion of the philtrum of the upper lip.

The secondary palate closes by the merging of palatinal shelves separating the nasal and oral cavities.

The concept presented above differs from the classic description of facial development based on so-called facial processes. According to this classic concept, there is a frontal process from which grow a medial and a lateral nasal processus around the olfactory placode transforming the placode into the groove. A globular (premaxillary) process, a derivative of the medial nasal process, reaches the border of the oral cavity. According to the classic concept, the nose is contributed to by the medial and lateral nasal processus, and the primary palate by the merging maxillary and globular processus.

The classic concept was criticized by Streeter (1948) as being oversimplified and incorrect. He pointed out that the prolongations do not have free ends (processus) that meet and fuse, nor is there any ectoderm absorbed over their surfaces. Streeter introduced a concept of facial swellings and ridges that corresponds to centers of growth in the underlying mesenchyme. The furrows between the centers smooth out as the mesenchyme proliferates. The secondary palate closes by the merging of palatinal shelves separating the nasal and oral cavities.

The teeth (Fig. 75c)

The teeth are derived phylogenetically from the skin in relation to the oral cavity. The ectodermal epithelium of the primitive upper and lower jaws thickens into the labiogingival ridges. A secondary ectodermal lamella, known as the dental lamina, grows from the labiogingival ridge into the alveolar primordia of the jaws. The epithelium of the labiogingival ridge disintegrates, separating the lips and cheeks, leaving the vestibule of the oral cavity. The dental lamina gives rise to tooth buds, which represent the primordia of the deciduous teeth. The

buds of definitive teeth grow by a secondary proliferation of the dental lamina and are located lingual to the deciduous primordia. The teeth buds transform into belly-like caps, or enamel organs, which detach from the surface epithelium, and each tooth cap becomes enclosed by a mesenchymal dental sac with a protruding dental papilla. The cells of the invaginated inner epithelium of the ectodermal tooth caps transform into ameloblasts and produce the prisms of enamel. The enamel is an intracellular product that is added to the apical cellular membrane. Mesenchymal cells of the dental papilla adjacent to the inner enamel epithelium become odontoblasts, which deposit predentin on the enamel. The predentin mineralizes and changes into dentin. The dentin of the root is deposited by odontoblasts onto a thin epithelial sheath formed by enamel epithelium inferior to the crown of the tooth. The mesenchymal cells of the dental sac change into cementoblasts, which cover the dentin and a narrow portion of adjacent enamel with cementum.

The hypophysis (Figs. 199–203 and 205)

The hypophysis is composed of two parts of different origin: the adenohypophysis originates from the ectodermal Rathke's pouch; the neurohypophysis develops from the neuroectodermal evagination from the bottom of the infundibular depression of the diencephalic vesicle. The Rathke's pouch detaches from the pharynx by the end of the second month. The lumen of the pouch is present in the adenohypophysis as a cleft between the distal and intermedial parts. The neurohypophysis represents terminals for the neuroendocrine hypothalamo-hypophyseal traits synthesizing vasopressin and oxytocin. The adenohypophysis contains different cells synthesizing tropic hormones such as somatotropin, prolactin, adrenocorticotropin, thyrotropin, follicule-stimulating hormone, and luteinizing hormone.

COLOR-PLATES

COLOR PLATE I: OOCYTE, BLASTOGENESIS – STAGES 1-0 TO 5-2.

Figure 1c. Unfertilized oocyte from a ripe Graafian follicle. The oocyte is enclosed within the zona pellucida covered by attached follicular cells. Irregular particles of condensed chromatin are present within the nucleus. Toluidine blue stain of a semi-thin section.

Figure 2c. Unfertilized oocyte with a detached first polar body within the zona pellucida. Some follicular cells are still attached to the zona. The perivitelline space underneath the zona becomes distinct. Toluidine blue stain.

Figure 3c. Oocyte with a detaching second polar body. Toluidine blue stain.

Figure 4c. Implanted bilaminar embryo with a disrupting primary yolk sac. Stage 4-2. Lacunae are present within the trophoblastic shell. Colloid-like substance filling the primary yolk sac mixes with the fluid present in the primary mesoderm. PAS–H stain.

Figure 4c. Transverse section of a bilaminar embryo. Stage 4-3. Amnionic sac and secondary yolk sac with embryonic disk inbetween. PAS–alkaline phophatase blue stain. Purple stain of endodermal cells is related to the presence of glycogen.

Figure 6c. Transverse section of an early trilaminar embryo exhibiting formation of the primitive node. Stage 5-1. Formation of the primitive node precedes formation of the primitive groove. H–E stain.

Figure 7c. Detail of the primitive node. Stage 5-1. The clear cytoplasm of the node cells is related to the presence of glycogen. H–E stain.

Figure 8c. Transverse section of an early Lieberkühn canal. Stage 5-1 H–E stain.

Figure 9c. Unfixed trilaminar embryo with a notochordal process photographed after opening of the chorion. Stage 5–2. The embryo attached to the chorion consists of an amniotic sac and a yolk sac (artificially opened and pushed to the right side. A primitive groove is present on the embryonic disk.

Figure 10c. Longitudinal section of the dissected embryo depicted in Figure 9c. The disrupted endodermal yolk sac is pushed to the right. The allantois evaginates into the connecting stalk.

Figure 11c. Cross section of a presomite trilaminar embryo with a chordomesodermal process through the primitive groove. Stage 5-2. Proliferation of undifferentiated mesoderm (mesoblast) from the ectoderm of the primitive groove into the space between the ectoderm and the endoderm of the embryonic disk. AP–H stain.

Figure 12c. Cross section of a presomite trilaminar embryo with a notochordal process through the neurenteric canal. Stage 5-2. Undifferentiated mesoderm (mesoblast) proliferates from lateral lips of the blastoporus (neurenteric canal). AP blue stain.

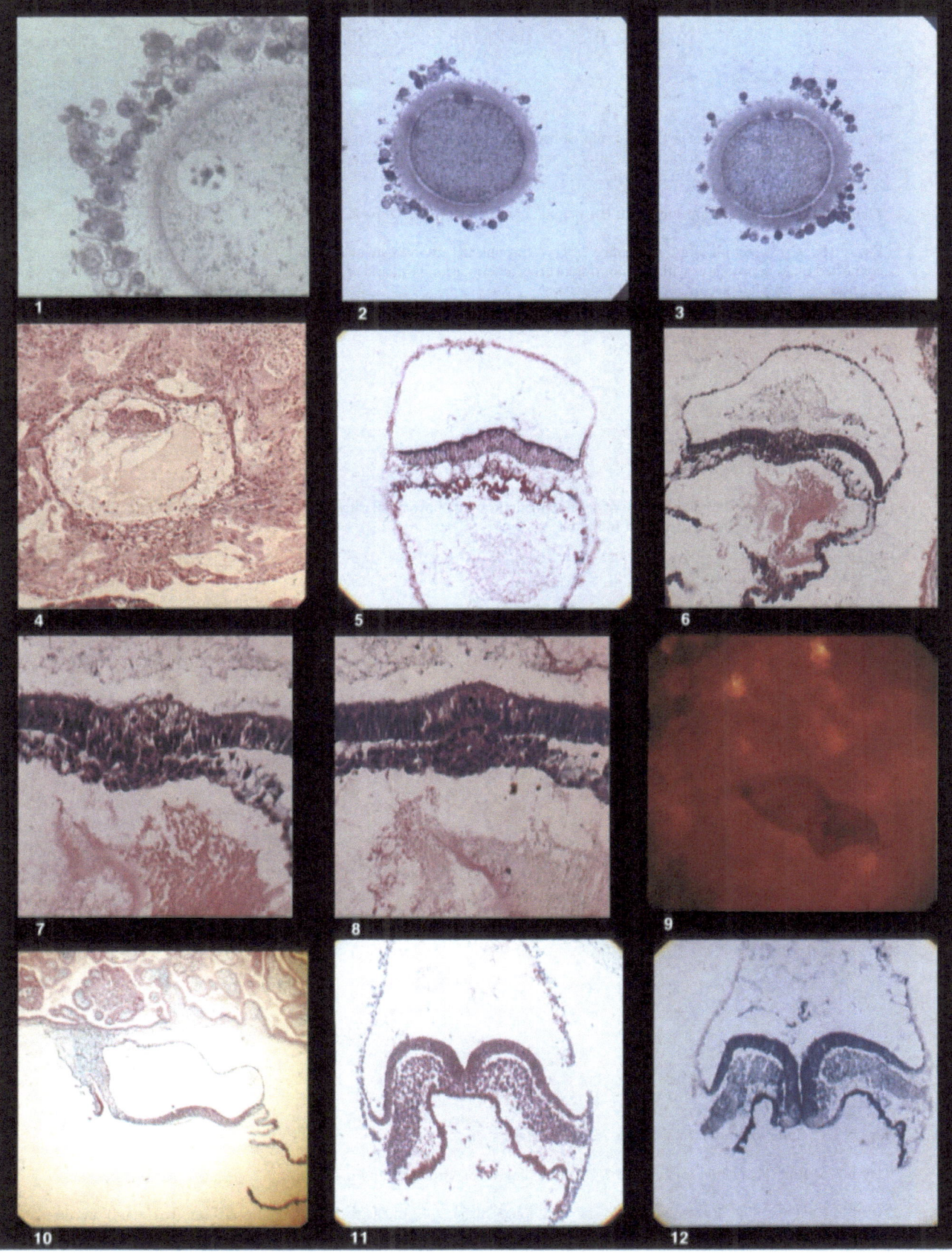

COLOR PLATE II: EMBRYOS – STAGES 6-2 TO 7-4

Figure 13c. Cross section of a three-somite embryo through the hindgut. Stage 6-1. The ventral portion of the neural groove is apposed to the notochordal plate. Undifferentiated mesoderm (mesoblast) is present between the ectoderm and the endoderm. Striking differences in AP activity (blue) in embryonic layers are evident. AP blue stain, carmin.

Figure 14c. Embryo 2.5 mm long with ten paired somites. Stage 6-2. Unfixed.

Figure 15c. Cross section of a three-somite embryo through the second pair of somites. Stage 6-1. Mesodermal cells of somites are connected by long processus with neighboring neuroectoderm, ectoderm, and endoderm. The ventral plate of the neural groove and the notochordal plate are in contact. AP–H stain.

Figure 16c. Embryo 1.5 mm long (stage of developing limbs, bud of proximal extremity), attached to the dissected chorion. Stage 7-1. AP surface stain, vessels injected with india ink.

Figure 17c. Semitransparent embryo 6.5 mm long. Stage 7-2. Note the optic cup, otocyst, pharyngeal arches and pouches, and proximal and distal limb buds. Unstained specimen.

Figure 18c. Embryo 7 mm long. Stage 7-2. Developing limbs; proximal and distal limb buds. AP surface stain.

Figure 19c. Embryo 6 mm long (proximal limb bud), attached to the dissected chorion. Stage 7–2. AP surface stain.

Figure 20c. Embryo 6 mm long. Stage 7-3. AP surface stain, semitransparent.

Figure 21c. Embryo 9 mm long. Stage 7-3. Developing limbs; proximal limb two segments; distal unsegmented bud. AP surface stain.

Figure 22c. Embryo 10 mm long. Stage 7-3. AP surface stain.

Figure 23c. Embryo 12 mm long. Stage 7-4. Developing limbs; proximal and distal limbs bisegmented. AP surface stain.

Figure 24c. Sagittally dissected embryo 11 mm long. Stage 7-4. Intensive alkaline phosphatase (red stain) in the CNS; glucoseaminoglycans (blue stain) are concentrated within condensed mesenchyme of the primitive vertebral column; and primitive valvular apparatus of the heart. AP–alcian blue stain.

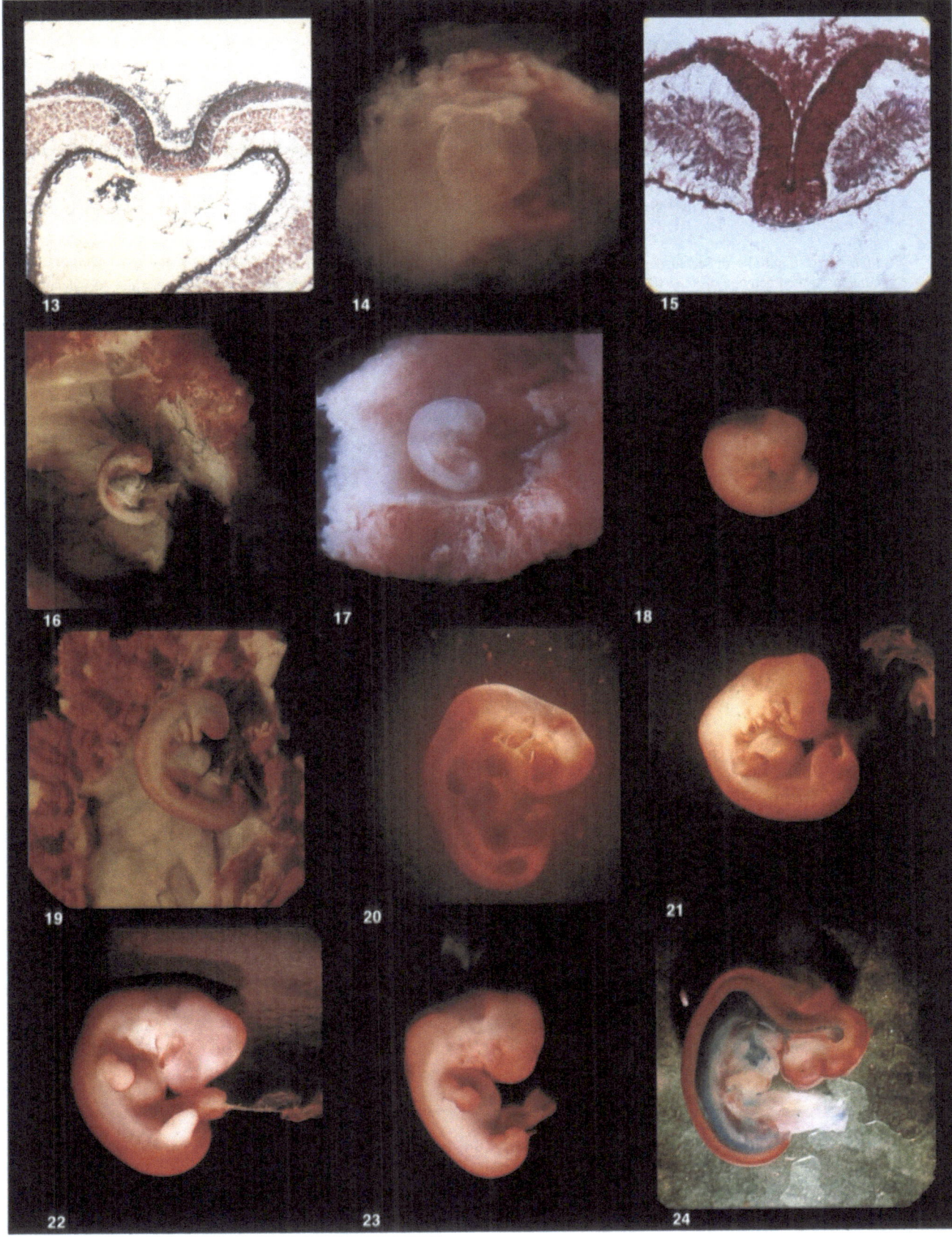

COLOR PLATE III: EMBRYOS – STAGES 7-5 TO 9.

Figure 25c. Embryo 12 mm long. Stage 7-4. Developing limbs; proximal and distal limbs bisegmented. The area of fourth cerebral ventricle is artificially damaged. Alcian blue stain.

Figure 26c. Embryo 14 mm long. Stage 7-5. Developing limbs; digital rays and foot plates. Artificial shrinkage of the head is evident. AP surface stain.

Figure 27c. Sagittally dissected embryo 18 mm long. Stage 7-5. The following structures are evident: brain and medullary tube, condensed mesenchyme of axial skeleton, heart, lung, liver, and umbilical cord with a physiologic umbilical hernia. AP–alcian blue stain.

Figure 28c. Embryo 18 mm long. Stage 7-6. Developing limbs; digital tubercles. AP surface stain.

Figure 29c. Embryo 24 mm long. Stage 8-1. Differentiated limbs; open eyes. AP surface stain.

Figure 30c. Embryo 20 mm long. Stage 8-1. Differentiated limbs; open eyes. The hands do not reach the midline. Anterior view. AP surface stain.

Figure 31c. Embryo 30 mm long. Stage 8-2. Fusing eyelids. AP surface stain.

Figure 32c. Fetus, 45 mm long, with fused eyelids from the tenth postconceptional week. Stage 9-1. AP surface stain.

Figure 33c. Unfixed and unstained fetus from the end of the third gestational month.

Figure 34c. Sagittally dissected embryo 24 mm long. Stage 8-1. Differentiated limbs; open eyes. AP-alcian blue stain.

Figure 35c. Sagittally dissected embryo 32 mm long. Stage 8-2. Fusing eyelids. Alcian blue stain.

Figure 36c. Sagittally dissected early fetus. Stage 9-1 (early fetus stage). AP-alcian blue stain.
On sagittal dissections, note the developmental changes of CNS, vertebral column, heart, liver, and intestines.

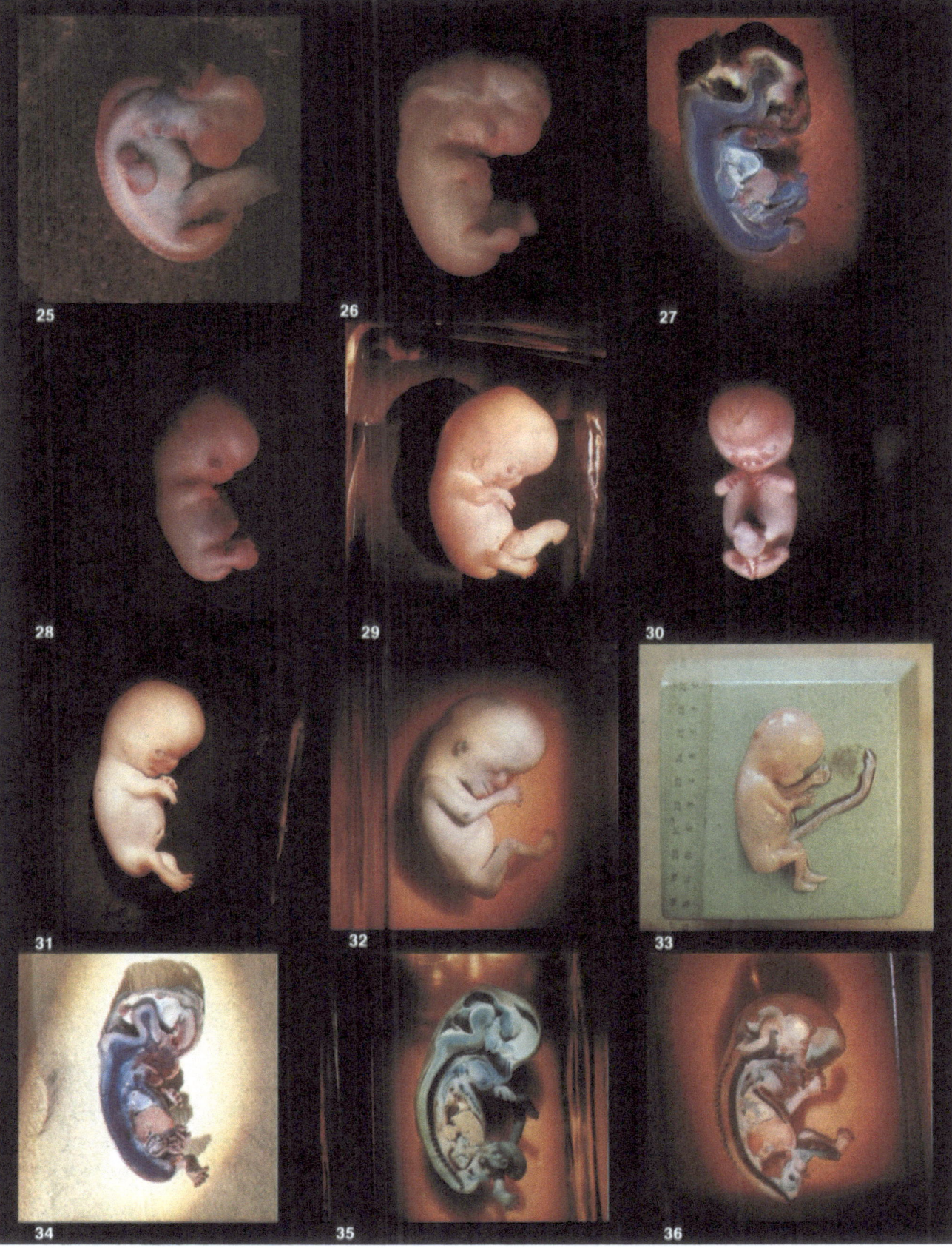

COLOR PLATE IV: EARLY PLACENTAL DEVELOPMENT (FIGS. 37C–39C), DEVELOPMENT OF SKELETON (FIGS. 40C–45C), AND DEVELOPMENT OF KIDNEYS (FIGS. 46C–48C)

Figure 37C. Trophoblastic shell with primary chorionic villi and empty, bloodless, trophoblastic lacunae representing the early intervillous space. Dark red staining of cytotrophoblasts is related to the presence of glycogen. Bilaminar embryonic primordium is present within the chorionic cavity. Stage 4-2. PAS–H stain.

Figure 38C. Chorion with secondary chorionic villi. No vascular primordia are present within the mesenchyme of villi. Stage 5-2. PAS–alcian blue stain.

Figure 39C. Chorion with tertiary chorionic villi. Endothelial cords and tubes with no blood cells present within the mesenchyme of villi. Maternal blood is evident in the intervillous space. AP–H stain.

Figure 40C. Cross section of an embryonic body showing a closed neural tube, a somite with a distinct myocoele with a myotome (dorsolateral portion), with a sclerotome (ventromedial portion spreading toward the glycogen rich notochord and to the aorta), and with a dermatome (ventrolateral portion). Adjacent to mesonephric blastema is a glycogen-rich primordium of the mesonephric (Wolffian) duct. Stage 6-3. AP blue stain, glycogen purple.

Figure 41C. Cartilaginous skeleton and brain (blue) in a transparent embryo. Stage 8-1. Alcian blue stain, cleared specimen.

Figure 42C. Cartilaginous skeleton of the upper limb, the base of skull, the cervical spine, and the thorax. Stage 8-2. Alcian blue stain, cleared specimen.

Figure 43C. Blastematous mesenchymal condensation representing skeletal primordia within a proximal limb and thorax. Stage 7-4. Alcian blue stain.

Figure 44C. Cartilaginous skeleton of the limbs, the thorax, and the spinal column. Stage 8-2. Alcian blue stain, cleared specimen.

Figure 45C. Ossification of the hand in a four-month-old fetus. Stage 9-2. Cartilage blue, bone red. Alizarin red, toluidine blue stain, cleared.

Figure 46C. Dissected retro- and intraperitoneal organs. Stage 8-1. Structures depicted: diaphragm, adrenals, kidneys, gonads (ovaries), and radix of mesentery. AP-alcian blue stain.

Figure 47C. Adrenals and kidneys. Stage 8-2. The right adrenal was removed, but the left one is intact. The right kidney was dissected. AP–alcian blue stain.

Figure 48C. Dissected kidney from an early fetal stage (9-1). Nephrons are red; collecting tubules, calyces, and pelvis are unstained. The mesenchyme of renal papillae is blue. AP–alcian blue stain.

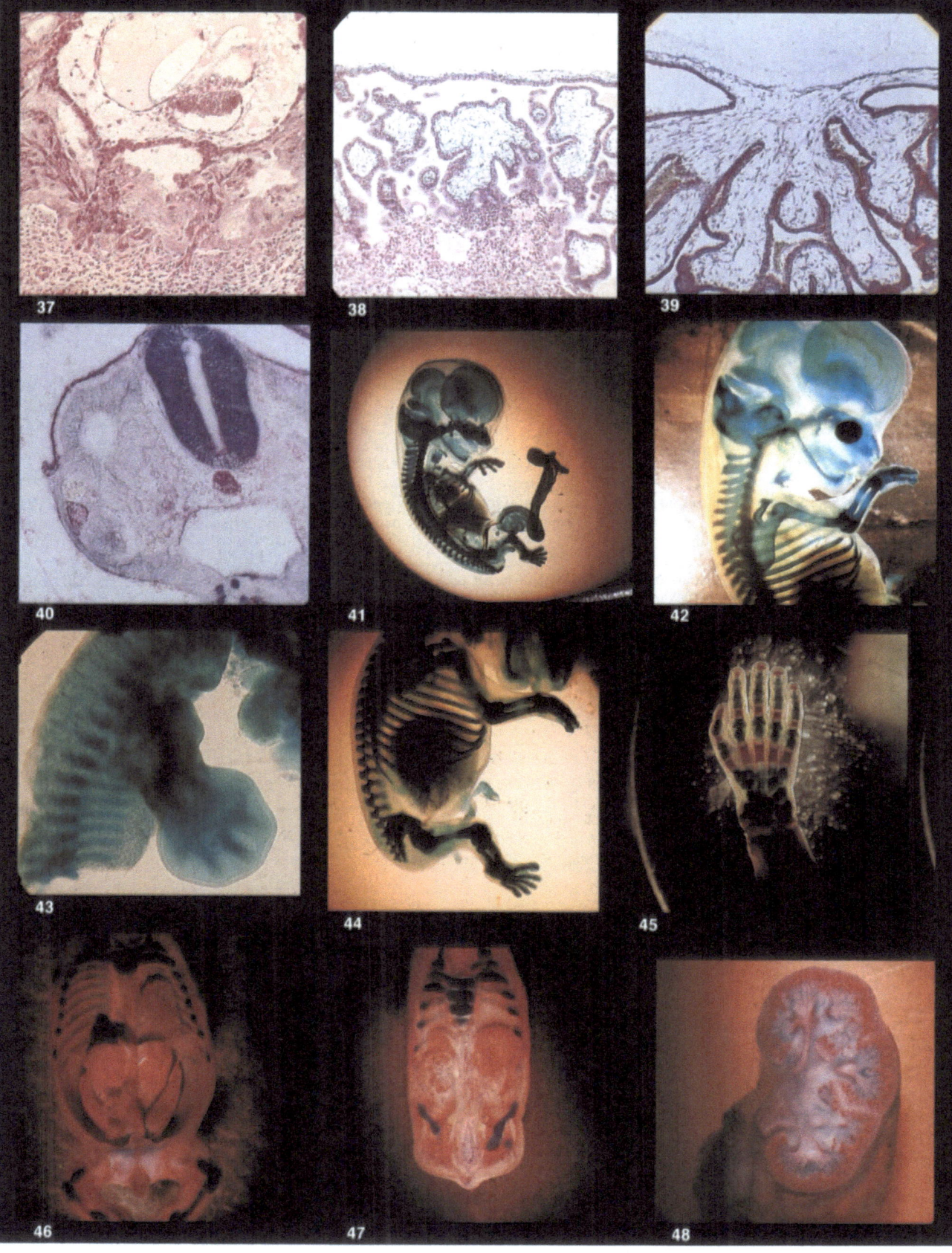

37

38

39

40

41

42

43

44

45

46

47

48

COLOR PLATE V: DEVELOPMENT OF THE GONADS

Figure 49C. Migrating primordial germ cells with long processus (pseudopodia) in a 12-mm-long embryo. Stage 7-4. The purple stain is due to glycogen content. PAS stain.

Figure 50C. Migrating primordial germ cells (blûe) in the gut endoderm and within the mesentery of a 3.5-mm-long embryo. Stage 6-3. AP blue stain.

Figure 51C. Primordial germ cells (red) concentrated in the medioventral portion of urogenital ridges of an 11-mm-long embryo. Stage 7-4.

Figure 52C. Cross section of the mesonephric and genital ridge in a 9-mm-long embryo. Stage 7-4. H-E–alcian blue stain.

Figure 53C. Section of the mesonephric and the genital ridge in a 15-mm-long embryo. Stage 7-6. The genital ridge exhibits an early testicular differentiation. Testicular germ cells temporarily lose alkaline phosphatase and glycogen coincidentally with formation of testicular cords. AP–H stain.

Figure 54C. Testis in a 22-mm-long embryo. Stage 8-1. AP positivity (and glycogen) has reappeared in the germ cells. Surface epithelium, testicular cords (delineated by basement membranes), and mesenchymal stroma are evident. AP–silver impregnation of reticulum.

Figure 55C. Testis of a 22-mm-long fetus. Stage 8-1. The testicular cords are composed of embryonic Sertoli cells and germ cells (dark red). The mesenchymal interstitium contains a specific glucoseaminoglycan (blue); connective tissue of the tunica albuginea is PAS positive, 19 mm embryo. PAS-alcian blue stain.

Figure 56C. Testis of a 45-mm-long fetus. Stage 9-1. The interstitium contains numerous epithelioid interstitial cells (of Leydig). Azan stain.

Figure 57C. Ovary in a 46, XX embryo, 21 mm long. The stage germ cells are red and the argyteophillic fibers are black. AP–silver impregnation of reticulum stain.

Figure 58C. Ovary of a 160-mm-long fetus. Alkaline-positive germ cells (oogonia) are present in the surface zone of the cortex. The deep cortical zone contains groups of oocytes entering the meiotic prophase. AP–H stain.

Figure 59C. Perinatal ovary from a newborn with compact singlelayered, multilayered, and vasicular follicles. Azan stain.

Figure 60C. Dissected abdominal cavity in an 18-mm-long embryo, revealing the topography of adrenals, kidneys, and genital ridges and ducts. Stage 7-6. AP surface stain.

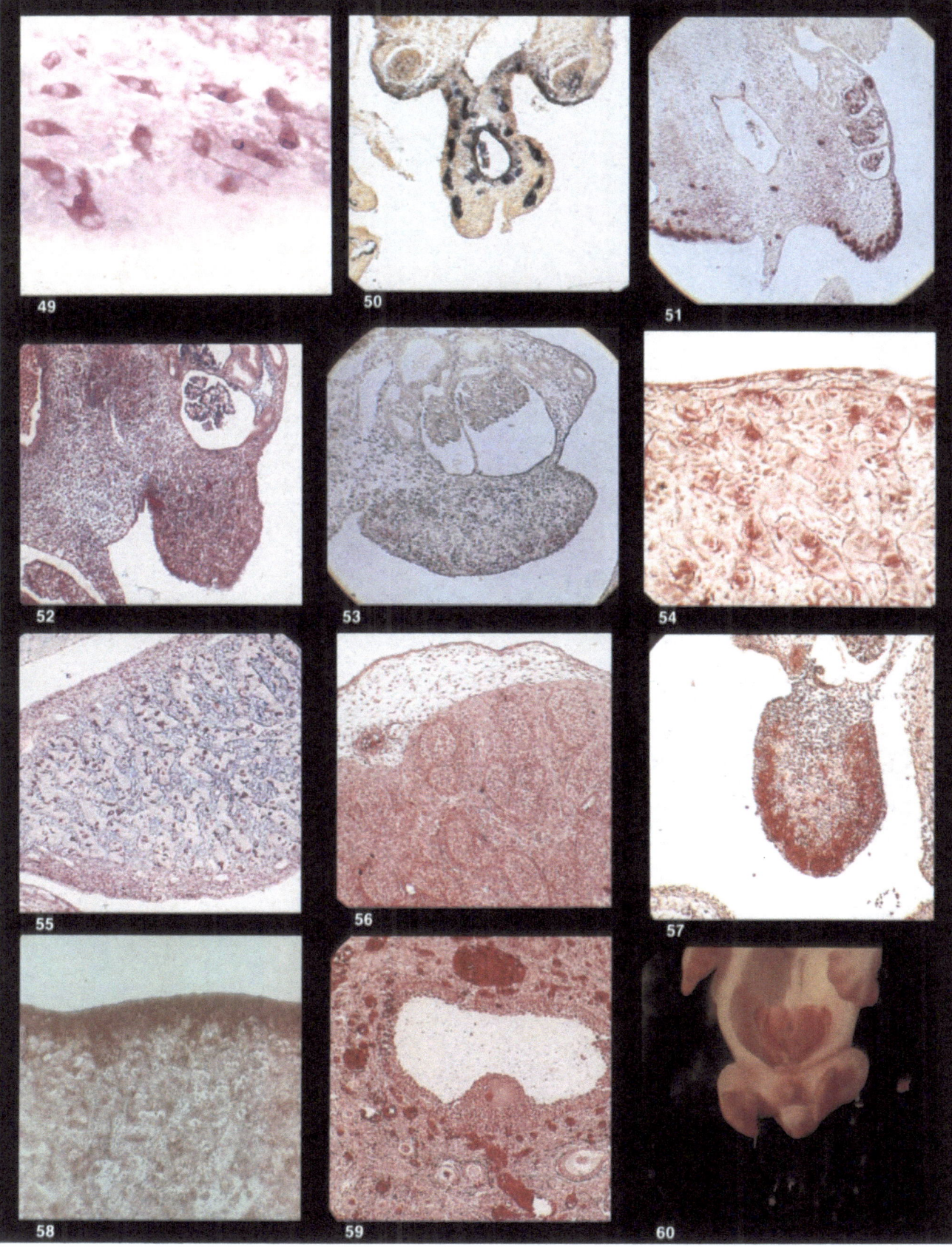

COLOR PLATE VI: DEVELOPMENT OF THE GENITAL DUCTS

Figure 61C. Genital ridges (ovaries) and fusing paramesonephric (Müllerian) ducts of a 24-mm-long embryo. Stage 8-1. AP surface stain – alcian blue.

Figure 62C. Dissected pelvis of a 35-mm-long embryo, showing regression of paramesonephric ducts. Stage 8-2. Regression is related to a glucoseaminoglycan diffusing from the testes into the caudal ligaments of mesonephric ridges (testicular gubernaculum). AP–alcian blue stain.

Figure 63C. Testes and regressed paramesonephric (Müllerian) ducts in a 40-mm-long fetus. AP–alcian blue stain.

Figure 64C. Adrenals, kidneys, ovaries, and fusing paramesonephric (Müllerian) ducts in a 30-mm-long embryo. Stage 8-2. AP surface stain–alcian blue.

Figure 65C. Sagittal pelvic dissection in a 150-mm-long female fetus. Structures depicted: symphysis, urinary bladder, female urethra, urogenital sinus, endodermal attachment of fused paramesonephric ducts, clitoris, rectum, and coccys. AP surface stain–alcian blue.

Figure 66C. Sagittal pelvic dissection in a 220-mm-long female fetus. Structures depicted: uterus, vagina, urinary bladder, female urethra, vulva, and rectum. AP surface stain–alcian blue.

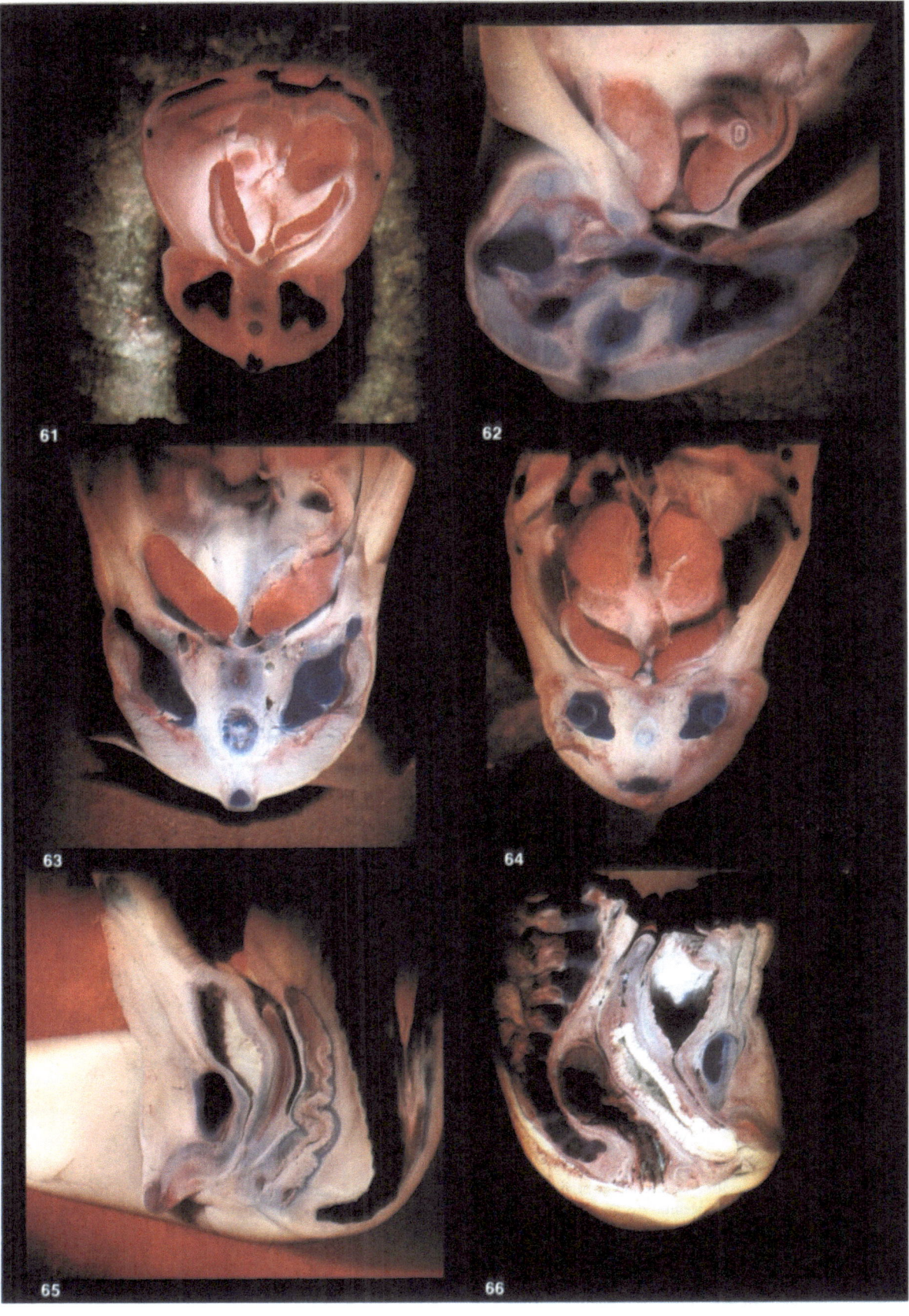

COLOR PLATE VII: DEVELOPMENT OF THE EXTERNAL GENITALIA – INDIFFERENT AND MALE EXTERNAL GENITALIA

Figure 67C. Primordia of external genitalia located between the umbilical cord and the tail in a 12-mm-long embryo. Stage 7-5. AP surface stain.

Figure 68C. Indifferent external genitalia of a 20-mm-long embryo. Stage 7-7. AP activity (red) is related to growth of the urethral plate. AP surface stain.

Figure 69C. Cross section of a closing urethral groove of a male fetus. Stage 9-1. AP (red) is located within the mesenchyme undergoing differentiation into the urethral spongy body. AP surface stain.

Figure 70C. Indifferent external genitalia of a 32-mm-long embryo. Stage 8-2. The phallus has a corporal portion and a glandular portion. Below the corporal portion is a distinct urethral groove. Labioscrotal swelling are located laterally to the phallus. Anal hillocks delineate the anal opening. AP surface stain.

Figure 71C. Masculinization of external genitalia in a 55-mm-long male fetus. The raphe of scrotum unifies the labioscrotal swellings into the scrotum. The anogenital distance increases. The incompletely closed cavernous urethra represents a 'physiologic' penile hypospadia.

Figure 72C. Parasagittally dissected penis of a 150-mm-long male fetus. The two corpora covernosa develop within the mesenchyme of the genital tubercle. The spongy body of the glans develops from the glandular portion, while the spongy body of the urethra originates from corporal portion of urethral plate. AP surface stain–alcian blue.

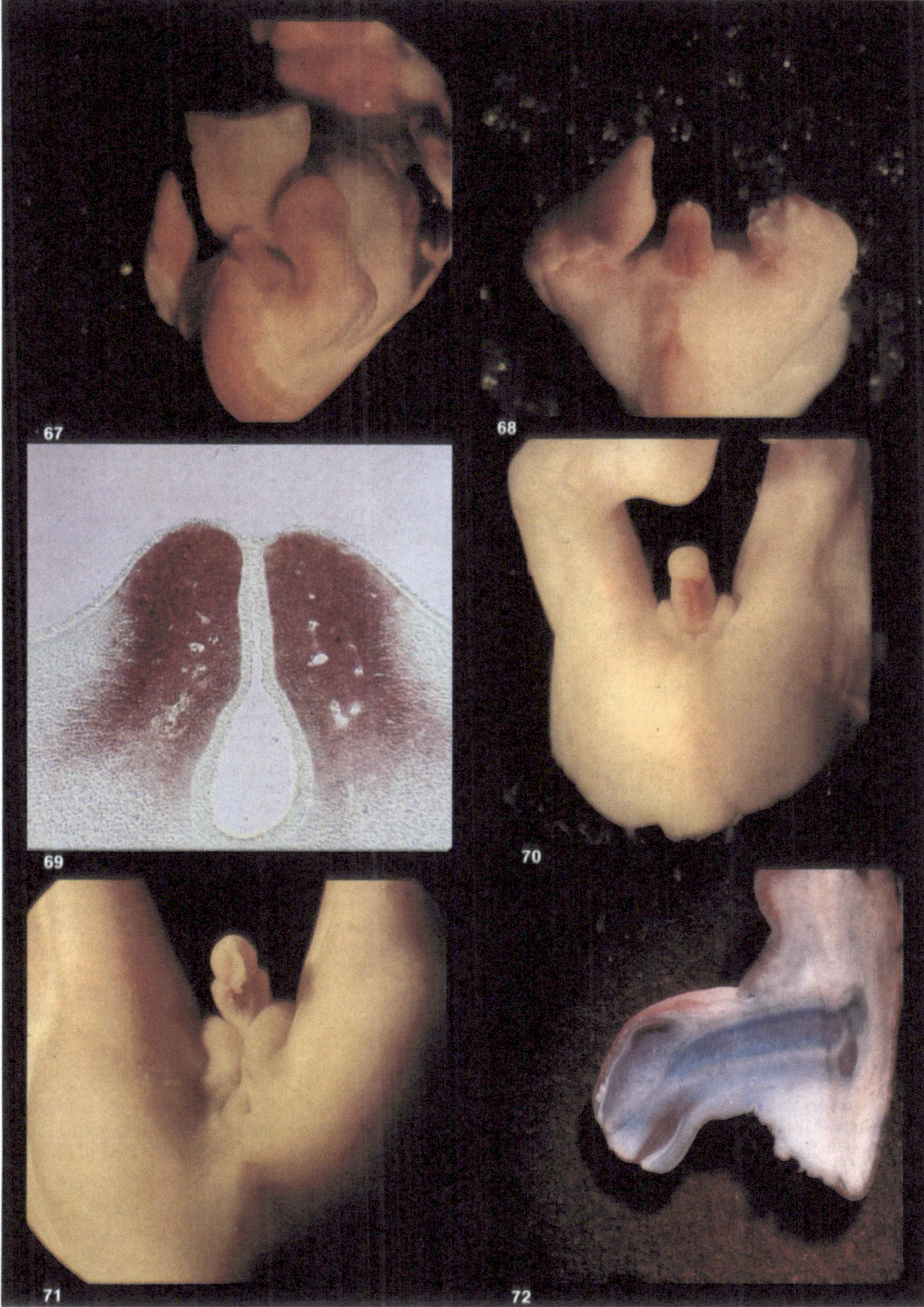

COLOR PLATE VIII: DEVELOPMENT OF THE FEMALE EXTERNAL GENITALIA, THE TONGUE AND THE LUNGS, AND THE BRAIN

Figure 73C. Feminization of external genitalia in a 38-mm-long female fetus. Stage 9-1. Neither the labioscrotal swellings nor the urethral groove fuse.

Figure 74C. Feminization of external genitalia in a 40-mm-long female fetus. Stage 9-1. Ventral flexion of phallus, rims of urethral groove changing into labia minora. Labioscoral swellings, future labia majora, remain in a lateral portion.

Figure 75C. Sagittal dissection of the lower jaw in a 200-mm-long fetus. Structures depicted: lower lip and chin with abundant hair, dental sac with a tooth cup, and dental papilla of a deciduous incisor within the alveolar portion of the jaw. AP–alcian blue stain.

Figure 76C. Tongue of a 200-mm-long fetus. Visualized structures: lingual corpus and radix. Circumvallate papillae are red. AP–alcian blue stain.

Figure 77C. Filiform papillae on the lingual apex.

Figure 78C. Circumvallate papillae anterior to the terminal sulcus stained by alkaline phosphatase.

Figure 79C. Dissected lung of a 220-mm-long fetus. Bronchi injected with red gelatin. Cleared specimen.

Figure 80C. Red-gelatin-injected specimen showing terminal bronchi branching into alveolar ducts and alveolar atria. Future alveolar spaces are located between alveolar ducts and lobular septa; 220-mm-long fetus; connective tissue visualized by hematoxylin stain, cleared specimen.

Figure 81C. Head of a 35-mm-long embryo. Stage 8-2. Structures depicted: brain and cartilaginous base of skull. Each cerebral hemisphere exhibits frontal, occipital, temporal, and olfactory lobes and an insula. The diencephalon evaginated into an optic recessus and the hypophysis located within the sella. The rhombencephalon is relatively large. The rhombencephalon consists of cerebellum, pons, and medulla oblongata. Alcian blue stained, methylsalicylate cleared specimen.

Figure 82C. Sagittally dissected head of a 28-mm-long embryo. Stage 8-1. Structures depicted: lateral ventricle communicating with the third ventricle through an interventricular foramen, third ventricle, a broad mesencephalic aqueduct, and a fourth ventricle. AP red surface stain.

Figure 83C. Transvere dissection of the brain of a 25-mm-long embryo. Stage 8-1. Dorsal view. Structures depicted: choroid plexuses in the lateral ventricles of hemispheres; corpora striata (pl. from corpus striatum) of hemispheres fusing with diencephalon. The medulla oblongata represents floor of the fourth ventricle.

Figure 84C. Dissected brain (upper portion) of the same specimen as in Figure 83C. Cerebral hemispheres, dissected diencephalon and metencephalon, rhombic lips fusing into a cerebellar plate. AP–alcian blue stain.

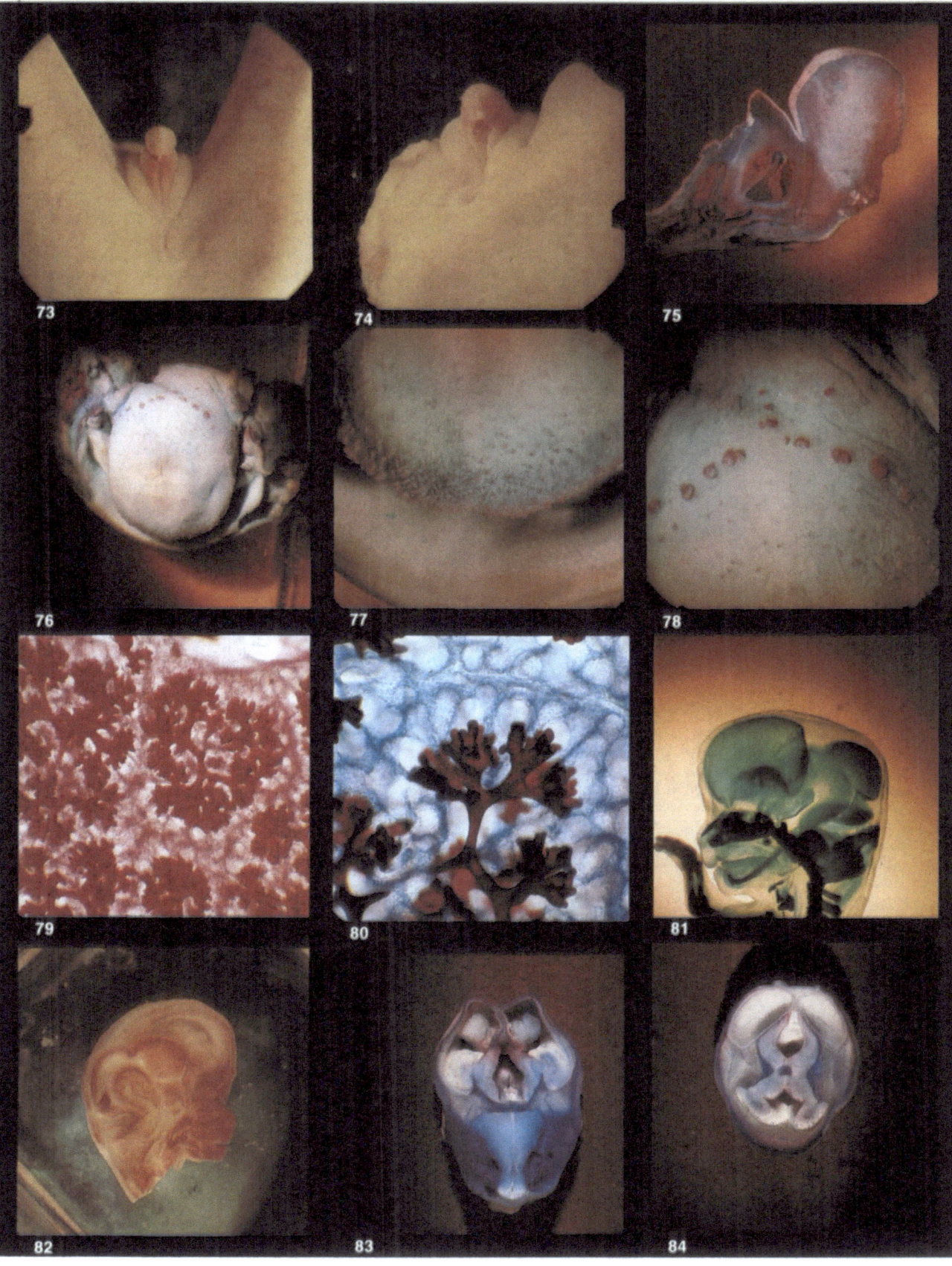

COLOR PLATE IX: DEVELOPMENT OF THE BRAIN, EYE, EAR, AND PALATE

Figure 85C. Dissected brain in situ in a 40-mm-long fetus. Stage 9-1. Structures depicted: cerebral hemispheres, chorioid plexus protruding into lateral ventricles, third ventricle, lateral diencephalic walls fused with the basal walls of the hemispheres – caudate nucleus and lentiform nucleus are evident. Cartilaginous internal ear is present lateral to mesencephalon. AP red surface stain–alcian blue.

Figure 86C. Transverse dissection of the head in a 40-mm-long fetus: cerebral hemispheres with lateral ventricles and choroid plexus, mesencephalon dissected transversely, and cerebellar plate with a vermis and hemispheres. AP red surface stain–alcian blue.

Figure 87C. Transverse histologic section of the head in an 8-mm-long embryo. Stage 7-3. Structures depicted: diencephalon extending laterally into optic cups, lens pits, olfactory placodes, and prosencephalon. Hypophyseal (Rathke) pouch apposed to the diencephalic floor. AP red–H stain.

Figure 88C. Section of optic cup with invaginated lens vesicle; a 10-mm-long embryo. Stage 7.3. Glucoseaminoglycans (blue) are present between the lens vesicle and the optic cup. H–alcian blue stain.

Figure 89C. Dissected eye of a 200-mm-long fetus. Remnants of vessels of the papillary membrane on the surface of lens; radial capillaries of the ciliary body. AP red stain–alcian blue.

Figure 90C. Transversely dissected head exposing oral cavity. Structures depicted: tongue, pharyngotympanic tubes, cartilaginous primordia of the minor ear, vestibular and cochlear ganglia, and medulla oblongata. AP red stain–alcian blue.

Figure 91C. Sagittal dissection of the head and trunk of an 18-mm-long embryo. Stage 7-7. Structures depicted: cartilage of the nasoseptal portion of nasodorsal center, upper lip, tongue interposed between palatinal shelves, lower jaw, heart, liver, and stomach. AP red stain–alcian blue.

Figure 92C. Frontal dissection of a head illustrating the mutual position of maxilla and nasal septum anteriorly to the apex of tongue; a 25-mm-long embryo. Stage 8-1.

Figure 93C. Frontal dissection of a head illustrating the positon of palatinal shelves and interposed tongue; a 25-mm-long embryo. Stage 8-1.

Figure 94C. Transversely dissected head through the oral cavity of a 24-mm-embryo. Stage 8-1. Structures depicted: upper lip, premaxilla fused with palatinal shelves, palatinal shelves, nasal septum, pharyngotympanic tubes, internal and midline ear enclosed within cartilaginous capsules, and pons. AP red stain–alcian blue.

Figure 95C. Dissected upper jaw with fusing palatinal shelves; a 32-mm-long embryo. Stage 8-2. Alcian blue (overstained).

Figure 96C. Transversely dissected head at stage 9-1. Structures depicted: upper lip, closed hard palate, pharyngeal lumen, and cartilages of petrous bones containing primordia of inner ear. AP red stain–alcian blue.

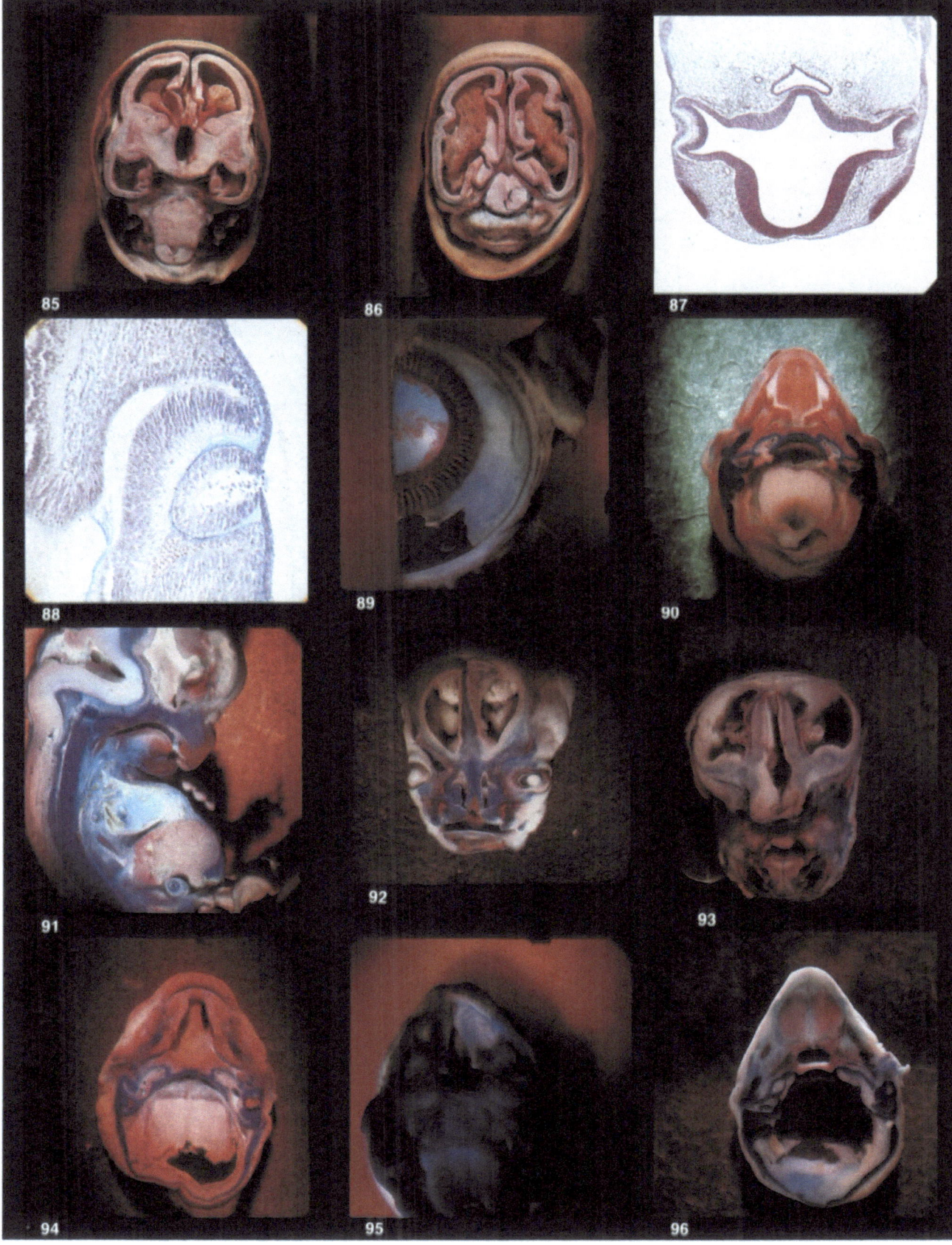

INDEX